PREPARATION OF
A FEASIBILITY STUDY FOR
NEW NUCLEAR POWER PROJECTS

The following States are Members of the International Atomic Energy Agency:

AFGHANISTAN
ALBANIA
ALGERIA
ANGOLA
ANTIGUA AND BARBUDA
ARGENTINA
ARMENIA
AUSTRALIA
AUSTRIA
AZERBAIJAN
BAHAMAS, THE
BAHRAIN
BANGLADESH
BARBADOS
BELARUS
BELGIUM
BELIZE
BENIN
BOLIVIA, PLURINATIONAL
  STATE OF
BOSNIA AND HERZEGOVINA
BOTSWANA
BRAZIL
BRUNEI DARUSSALAM
BULGARIA
BURKINA FASO
BURUNDI
CABO VERDE
CAMBODIA
CAMEROON
CANADA
CENTRAL AFRICAN
  REPUBLIC
CHAD
CHILE
CHINA
COLOMBIA
COMOROS
CONGO
COOK ISLANDS
COSTA RICA
CÔTE D'IVOIRE
CROATIA
CUBA
CYPRUS
CZECH REPUBLIC
DEMOCRATIC REPUBLIC
  OF THE CONGO
DENMARK
DJIBOUTI
DOMINICA
DOMINICAN REPUBLIC
ECUADOR
EGYPT
EL SALVADOR
ERITREA
ESTONIA
ESWATINI
ETHIOPIA
FIJI
FINLAND
FRANCE
GABON
GAMBIA, THE

GEORGIA
GERMANY
GHANA
GREECE
GRENADA
GUATEMALA
GUINEA
GUYANA
HAITI
HOLY SEE
HONDURAS
HUNGARY
ICELAND
INDIA
INDONESIA
IRAN, ISLAMIC REPUBLIC OF
IRAQ
IRELAND
ISRAEL
ITALY
JAMAICA
JAPAN
JORDAN
KAZAKHSTAN
KENYA
KOREA, REPUBLIC OF
KUWAIT
KYRGYZSTAN
LAO PEOPLE'S DEMOCRATIC
  REPUBLIC
LATVIA
LEBANON
LESOTHO
LIBERIA
LIBYA
LIECHTENSTEIN
LITHUANIA
LUXEMBOURG
MADAGASCAR
MALAWI
MALAYSIA
MALI
MALTA
MARSHALL ISLANDS
MAURITANIA
MAURITIUS
MEXICO
MONACO
MONGOLIA
MONTENEGRO
MOROCCO
MOZAMBIQUE
MYANMAR
NAMIBIA
NEPAL
NETHERLANDS,
  KINGDOM OF THE
NEW ZEALAND
NICARAGUA
NIGER
NIGERIA
NORTH MACEDONIA
NORWAY
OMAN

PAKISTAN
PALAU
PANAMA
PAPUA NEW GUINEA
PARAGUAY
PERU
PHILIPPINES
POLAND
PORTUGAL
QATAR
REPUBLIC OF MOLDOVA
ROMANIA
RUSSIAN FEDERATION
RWANDA
SAINT KITTS AND NEVIS
SAINT LUCIA
SAINT VINCENT AND
  THE GRENADINES
SAMOA
SAN MARINO
SAUDI ARABIA
SENEGAL
SERBIA
SEYCHELLES
SIERRA LEONE
SINGAPORE
SLOVAKIA
SLOVENIA
SOMALIA
SOUTH AFRICA
SPAIN
SRI LANKA
SUDAN
SWEDEN
SWITZERLAND
SYRIAN ARAB REPUBLIC
TAJIKISTAN
THAILAND
TOGO
TONGA
TRINIDAD AND TOBAGO
TUNISIA
TÜRKİYE
TURKMENISTAN
UGANDA
UKRAINE
UNITED ARAB EMIRATES
UNITED KINGDOM OF
  GREAT BRITAIN AND
  NORTHERN IRELAND
UNITED REPUBLIC OF TANZANIA
UNITED STATES OF AMERICA
URUGUAY
UZBEKISTAN
VANUATU
VENEZUELA, BOLIVARIAN
  REPUBLIC OF
VIET NAM
YEMEN
ZAMBIA
ZIMBABWE

The Agency's Statute was approved on 23 October 1956 by the Conference on the Statute of the IAEA held at United Nations Headquarters, New York; it entered into force on 29 July 1957. The Headquarters of the Agency are situated in Vienna. Its principal objective is "to accelerate and enlarge the contribution of atomic energy to peace, health and prosperity throughout the world".

IAEA NUCLEAR ENERGY SERIES No. NG-T-3.3 (Rev. 1)

PREPARATION OF A FEASIBILITY STUDY FOR NEW NUCLEAR POWER PROJECTS

INTERNATIONAL ATOMIC ENERGY AGENCY
VIENNA, 2026

COPYRIGHT NOTICE

© IAEA, 2026

Printed by the IAEA in Austria
January 2026
STI/PUB/2130
https://doi.org/10.61092/iaea.cx54-5gec

IAEA Library Cataloguing in Publication Data

Names: International Atomic Energy Agency.

Title: Preparation of a feasibility study for new nuclear power projects / International Atomic Energy Agency.

Description: Vienna : International Atomic Energy Agency, 2026. | Series: IAEA nuclear energy series, ISSN 1995-7807 ; no. NG-T-3.3 (rev. 1) | Includes bibliographical references.

Identifiers: IAEAL 26-01806 | ISBN 978-92-0-126425-1 (paperback : alk. paper) | ISBN 978-92-0-126525-8 (pdf) | ISBN 978-92-0-126625-5 (epub)

Subjects: LCSH: Nuclear power plants — Planning. | Nuclear power plants — Economic aspects. | Nuclear power plants — Social aspects. | Nuclear power plants — Management.

Classification: UDC 621.039.5:620.9 | STI/PUB/2130

FOREWORD

The IAEA's statutory role is to "seek to accelerate and enlarge the contribution of atomic energy to peace, health and prosperity throughout the world". Among other functions, the IAEA is authorized to "foster the exchange of scientific and technical information on peaceful uses of atomic energy". One way this is achieved is through a range of technical publications including the IAEA Nuclear Energy Series.

The IAEA Nuclear Energy Series comprises publications designed to further the use of nuclear technologies in support of sustainable development, to advance nuclear science and technology, catalyse innovation and build capacity to support the existing and expanded use of nuclear power and nuclear science applications. The publications include information covering all policy, technological and management aspects of the definition and implementation of activities involving the peaceful use of nuclear technology. While the guidance provided in IAEA Nuclear Energy Series publications does not constitute Member States' consensus, it has undergone internal peer review and been made available to Member States for comment prior to publication.

The IAEA safety standards establish fundamental principles, requirements and recommendations to ensure nuclear safety and serve as a global reference for protecting people and the environment from harmful effects of ionizing radiation.

When IAEA Nuclear Energy Series publications address safety, it is ensured that the IAEA safety standards are referred to as the current boundary conditions for the application of nuclear technology.

In the context of increasing global energy demand, concerns about climate change and the need for energy security, many countries are considering or expanding their nuclear power programmes. This publication provides information on the preparation of a feasibility study for a new nuclear power project. The feasibility study serves as a critical tool for decision makers, providing the necessary information to make informed choices about the development of nuclear power infrastructure.

The publication is structured to guide the owner/operator through the processes critical to the successful implementation of a nuclear power plant. It addresses the necessary infrastructure, legal and regulatory frameworks; site selection and environmental impact assessments; and the integration of the nuclear power plant into the national grid. The publication also emphasizes the importance of stakeholder engagement, risk management and the development of a skilled workforce to support the project. It discusses prerequisites, organizational roles and responsibilities in conducting the feasibility study, and describes the contents of a feasibility study, providing detailed information on legal frameworks, electrical system analysis, site evaluation, environmental impacts, technology assessment, cost estimation, project implementation, risk management, staffing and national participation.

The IAEA wishes to thank all contributors to the drafting and review of this publication. The IAEA officers responsible for this publication were E. Mathet and A. Dutta-Ray of the Division of Nuclear Power and P. Dieguez-Porras of the Division of Planning, Information and Knowledge Management.

CONTENTS

1. INTRODUCTION

1.1. BACKGROUND

In the context of increasing energy demands to fuel economic growth and development, concerns about climate change and energy security, and in consideration of the improved safety and performance records of nuclear power plants (NPPs)[1], many countries have expressed interest in or are already implementing their first nuclear power programmes. In addition, several countries with operational NPPs are in the process of expanding their current nuclear power generating capacity.

To develop a successful NPP project, a country will first need to establish the necessary infrastructure required by a nuclear power programme. Reference [1] provides guidance for this infrastructure and the relevant activities.

According to Ref. [1], the successful implementation of a new nuclear power programme and project is carried out in three phases and addresses 19 specific issues related to the national nuclear infrastructure during each phase. Each phase ends with the achievement of a specific programme milestone, namely:

— Milestone 1: Ready to make a knowledgeable commitment to a nuclear power programme;
— Milestone 2: Ready to invite bids/negotiate a contract for the first NPP;
— Milestone 3: Ready to operate the first NPP.

A schematic representation of the phases and milestones is shown in Fig. 1 [1].

During Phase 1, various studies, including a pre-feasibility study (pre-FS), are conducted to support the development of the comprehensive report needed to support a knowledgeable decision by the government to commit to a nuclear power programme.

Once the country nuclear programme progresses to Phase 2 of the Milestones approach, it is expected that the project owner (owner/operator) will develop a feasibility study (FS) to define the framework and business case for successful project implementation and to establish the basis for the final investment decision to be taken by the project owner and to strengthen stakeholders' support. The FS also provides inputs to the development of the bid invitation specification (BIS) or to the technical specifications that the owner/operator issues to the vendor [2]. Therefore, the FS development is one of the critical processes to be completed during Phase 2, as presented in Fig. 1. There is no specific mention of the FS in Ref. [1]. Key elements of the FS are presented under Phase 2 — Management [1] and detailed for each relevant issue. In Ref. [3], activities are organized by infrastructure issues.

In 2014, this Technical Report[2] was published within the IAEA mandate to support Member States in their activities and to provide guidance on planning and delivering a project FS as part of the implementation of a nuclear power programme, thereby allowing the owner/operator to make informed decisions regarding the possible approaches to implement the NPP construction project.

The present revision considers the most recent experience of embarking and expanding countries in developing a FS and aligns these efforts with the latest practices and requirements for effective implementation of the Milestones approach in the development of national nuclear infrastructure.

---

[1] A list of the abbreviations used in the text is given at the end of the publication.

[2] INTERNATIONAL ATOMIC ENERGY AGENCY, Preparation of a Feasibility Study for New Nuclear Power Projects, IAEA Nuclear Energy Series No. NG-T-3.3, IAEA, Vienna (2014).

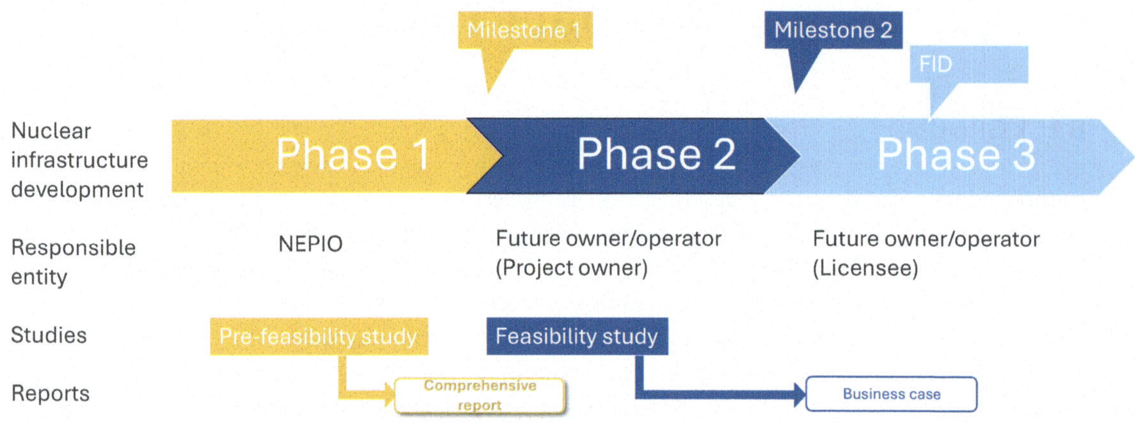

FIG. 1. Feasibility study development process and the Milestones approach. FID: final investment decision; NEPIO: nuclear energy programme implementing organization.

1.2. OBJECTIVE

This publication addresses all relevant issues related to the preparation of an FS for a new NPP project and describes good practices drawn from international experience. The main objectives of this publication are to:

— Provide guidance on the preparation of an FS to support the owner/operator decision making process for the project;
— Define an approach for the owner/operator to assemble the FS for the project, with particular emphasis on the specific site, environment, electrical grid integration issues, project structuring, management, budget and financing strategies;
— Present guidelines for implementation of the recommendations per Ref. [1].

The FS development process described in this publication is not constrained regarding the intended nuclear power project technology, design, size, location and industrial application, but the approach will vary depending on the specific project objectives and drivers. Guidance and recommendations provided here in relation to identified good practices represent expert opinion but are not made on the basis of a consensus of all Member States.

1.3. SCOPE

This publication describes the various steps generally undertaken to prepare the FS report in relation to the prerequisites of a nuclear power project considering that:

(a) A pre-FS is typically prepared by the nuclear energy programme implementing organization (NEPIO) [4] during Phase 1 of the IAEA Milestones approach to support the development of the comprehensive report to achieve Milestone 1 ("ready to make a knowledgeable commitment to a nuclear power programme). The pre-FS is at the programme level.
(b) The FS is prepared during Phase 2 of the IAEA Milestones approach to support the achievement of Milestone 2 (ready to invite bids/negotiate a contract for the first NPP). The FS builds upon the preliminary assumptions and conclusions of the pre-FS and the comprehensive report from the point of view of the plant owner/operator. The FS is at the project level.

1.4. STRUCTURE

This publication consists of the following sections:

— Section 1 includes background information and establishes an understanding of the place of FSs within the three phases of the Milestones approach.
— Section 2 presents the arrangements for the development of a project specific FS for an NPP, including the organization of the process, the main assumptions and preconditions, the project stakeholders, and the goals and objectives.
— Section 3 contains guidance for the preparation of a project specific FS organized according to its content. This consists of technical guidance for each topic, followed by suggestions concerning the content and specific issues to be considered while performing the evaluation for each topic. Separate subsections cover the project regulatory framework, integration of the plant into the electrical grid, and site and environmental issues. Nuclear specific issues are covered by the subsections dealing with the NPP technology and fuel cycle. Other subsections are dedicated to project implementation arrangements, risk management, human resources, national participation, economic and financial analysis, and the owner's specifications and requirements.
— Section 4 contains the conclusions.

This publication uses the phrase 'bid invitation specifications', which is applicable to a country using a competitive bidding process for the procurement of NPP technology. A country using an intergovernmental agreement, strategic partner or sole supplier, instead of a competitive process, therefore needs to interpret 'bid invitation specifications' as 'specifications for negotiating with the sole supplier'. In the Milestones approach [1], there is no distinction between the owner of the assets and the operator of the NPP, hence the term 'owner/operator' is used in this publication. In Phase 2, the owner/operator could also be a project management organization, and its structures would evolve in subsequent phases of the project.

1.5. USERS

The following organizations are the intended users of this publication:

— Owners or prospective owners and operators or organizations responsible for developing the NPP project;
— Government organizations, including the NEPIO;
— Engineering and consultancy service providers engaged in the implementation of specific programme development activities and studies or developing different level FSs.

The following organizations may also benefit from using this publication as a reference:

— Companies including technology providers, main contractors, manufacturers, service providers and others who are part of the new nuclear construction project supply chain;
— Regulatory bodies;
— Technical support organizations (TSOs);
— Any other key stakeholders in the nuclear power programme.

2. ARRANGEMENTS FOR A FEASIBILITY STUDY

2.1. PREREQUISITES

An FS for an NPP project is usually undertaken in the context of the development of a new nuclear programme or the expansion of an existing programme. It may be preceded by a pre-FS carried out to facilitate decisions during Phase 1 of programme implementation. The context in which an FS is undertaken is described in Ref. [1].

It is assumed that Phase 1 begins when the country's national energy planning has the potential to feature nuclear power in its energy mix. Usually, one of the first actions that the government takes in Phase 1 is the establishment of a NEPIO to coordinate the work needed to support a knowledgeable decision on the introduction of a nuclear power programme. The functions and responsibilities of the NEPIO are described in Ref. [4].

The main objectives of the NEPIO are to ensure the coordination and engagement of all important parties in the evaluation of the issues related to embarking on a nuclear power programme. As such, several investigations and studies are conducted and compiled to support a knowledgeable policy decision on whether to proceed with nuclear power. The key output of the NEPIO's work at the end of Phase 1 is a comprehensive report summarizing the results of the energy planning studies and the pre-FSs regarding the nuclear power programme. Should the NEPIO's report recommend proceeding with a nuclear power programme, it would be expected to also provide a plan or roadmap for the next steps of the programme.

When a decision is made to proceed with an NPP project, a project specific FS is undertaken in Phase 2 to comprehensively assess the project's feasibility. This is the ideal window of time in which to detect any underlying issue(s) that may challenge the project.

The project specific FS is typically carried out by the organization that is responsible for developing the NPP project through Phase 2. This organization often (but not always) subsequently becomes the NPP owner and/or operator.

The results of activities undertaken during Phase 1 to support the decision to proceed with the nuclear power option, including any pre-FSs, provide inputs for the preparation of the project specific FS. These may include:

— Identification of the legislative framework required for the nuclear power programme and demonstration of the capability to develop and promulgate the necessary laws;
— Identification of already issued regulations about the use of nuclear materials and definition of the applicable codes and standards to support the initial site selection and the design, licensing, construction, commissioning, operation and decommissioning of the NPP;
— Identification of the existing requirements governing a nuclear regulatory body and definition of a regulatory framework, which may include regulatory functions in the areas of licensing, review, assessment, inspection, enforcement and public information;
— Identification of the NPP owner and/or future NPP operator and engagement with the main stakeholders;
— Definition of the management model;
— Definition of the preferred contracting model(s);
— Inclusion in the government programme and budget allocation for the development, within an appropriate timeline, of the essential legal and safety programmes.

In addition, it is expected that the government will have committed to adhering to the necessary international legal instruments covering nuclear safety, security, safeguards and liability, and that it will be fostering international cooperation to support the programme.

Achieving the necessary level of technical and institutional competence will entail that substantial work be undertaken in Phase 2, requiring significant, continuing commitment from the government. It is assumed that the work of all organizations continues to be coordinated by the NEPIO towards developing the necessary infrastructure covering all 19 infrastructure issues and that an effectively independent regulatory body will be established that is capable of fulfilling all its functions and responsibilities.

It is also expected that for the purpose of the FS development, detailed vendor information and site specific data will be available or will be collected during the process. Many embarking countries may not have all the inputs necessary to undertake the FS, nor may all the necessary conditions be present for a first NPP. However, the study can still be conducted with particular attention paid to documenting the assumptions made, identifying the existing gaps and by providing a plan to address those gaps. In situations where certain prerequisites are not available as expected per the Milestones approach, the owner/operator agrees with the NEPIO or the regulatory body on reasonable assumptions to proceed with the FS, with actions to validate these assumptions once the prerequisite information becomes available.

Phase 2 usually concludes with the publication of the BIS or the specifications of the owner/operator. The FS that precedes the BIS development typically includes requirements and conditions that support drafting of the BIS, as described in Ref. [2]. Therefore, the project specific FS is a necessary preparatory step prior to any binding commitment with the potential engineering, procurement and construction (EPC) contractors, which elaborates the project requirements, structuring and implementation issues discussed in this publication.

After the bidding process has been completed, a technology and a vendor have been selected, and contract negotiations are underway, additional information about these aspects of the project will become available to update the FS to support the final investment decision to proceed with the procurement.

2.2. ORGANIZING THE FEASIBILITY STUDY

2.2.1. Roles and responsibilities of the owner/operator

Depending on the country's legislation, the owner/operator could be an organization such as a private utility or a partnership of such utilities. It could also be a government owned entity or a specially created project management organization. Regardless of the organizational model, the owner/operator of the future NPP is designated at the onset of Phase 2 of the nuclear power programme, as indicated in Ref. [1], and is responsible for managing the overall FS development. This includes the availability of adequate resources with basic knowledge about the FS development process, the requirements for execution of an FS and experience in evaluation of the FS report from a technical, economic and legal perspective. The owner/operator also needs to recruit the staff required to support the FS process with expertise in the following areas:

— Nuclear reactor technologies;
— Nuclear fuel cycle and radioactive waste management;
— Project and management systems, to manage the FS preparation process;
— Infrastructure in the country and at the future NPP site;
— Regulatory requirements;
— Legal and business aspects of FS preparation and review;
— Finance, to develop and analyse financing plans;
— Stakeholder communication and public information;
— Environmental impact assessment and energy matters.

Inputs from the proposed reactor technology provider(s) may be used to inform the development of the FS with respect to technology, organization, schedule and cost.

It is a common practice and advisable for the owner/operator to obtain assistance from well qualified external resources, such as a TSO or consultants [5] with experience in FS preparation and with specialized knowledge that may be lacking in the owner/operator organization. Such resources, however, take an advisory role to the owner/operator, who keeps full responsibility for the FS.

An external resource for the FS preparation could be necessary for the following reasons:

— Preparation of the FS requires a strong background in the technical, licensing, management, economic and financial aspects of the NPP project.
— The perceived objectivity of the evaluation is an important factor for the credibility of the FS for potential financial lenders and other interested parties.
— It is important to hire a consultant with no formal ties to the project, equipment manufacturers or marketers, so that an unbiased evaluation of the NPP project's operating potential and efficiency can be made.

Typically, the external resource is selected through a competitive process. It is good practice to establish a short list of three to six candidates from which to select. Some of the key factors to be considered in the selection of a consultant are:

— Relevant experience on related projects;
— Level of experience of the key personnel, such as project management and senior staff;
— Access to support resources;
— Past performance on client contracts;
— Meaningful partnerships/associations with local companies;
— Established quality management systems.

External resources may be requested to prepare specific chapters on behalf of the owner/operator that require very specific skills or qualifications, such as those relating to environmental marine fauna preservation.

As an important part of the FS development, it is expected that the main project stakeholders will be identified, with the importance of their engagement well understood, forming the basis for the respective strategy. It is expected that in Phase 2 the owner/operator will establish a stakeholder involvement strategy and plan and initiate engagement with identified stakeholders, as discussed in Ref. [3]. This strategy and plan may benefit from previous work performed at the national level in Phase 1 and is coordinated with the plans of the other key organizations through the NEPIO. The development of a stakeholder engagement strategy and plan supports the owner/operator's efforts in communicating and engaging with all identified interested parties. Reference [6] provides guidance on developing and implementing stakeholder engagement programmes and activities.

The owner/operator's area of responsibility includes necessary contacts with all related stakeholders prior to and during the FS preparation process. Communication with stakeholders offers a two way benefit:

— The owner/operator receives the information and guidance needed to prepare the FS (e.g. from the nuclear safety regulatory body for the licensing process and specific regulations, from the national electricity utility for the integration of the NPP into the national grid, from the environmental agency for the licensing process and specific regulations, from the Ministry of Education for human resources development).
— The other stakeholders are better prepared for measures to be taken in the later phases of the NPP project.

2.2.2. Roles and responsibilities of other organizations

In addition to the owner/operator, the government and/or NEPIO will provide contributions to the development of the FS. This information may be related to national infrastructure, such as the industrial capabilities for nuclear and/or non-nuclear safety related items or the education and training centres needed to develop the workforce. Some of this information may be available from pre-FSs that were conducted in Phase 1 but will likely need to be reviewed and updated before being incorporated. Similarly, the regulatory body, once established, provides the regulatory requirements for licensing an NPP, which is discussed in more detail in Section 3.2.2.

3. CONTENTS OF A FEASIBILITY STUDY FOR A NEW NUCLEAR POWER PLANT PROJECT

3.1. GENERAL OBSERVATIONS ON THE CONTENTS

As indicated in Section 1, any pre-FS conducted in Phase 1 provides information that is included in the comprehensive report to inform the decision to embark on a nuclear power programme. The FS, in comparison, supports the investment decision for one or more NPP projects to be integrated into the overall mix of electrical power generating plants. Some issues included in the FS may have been addressed in separate, standalone documents. In such cases, the FS needs to refer to these documents and, if necessary, to review, amend or update their contents, or simply summarize their findings and recommendations if the information is still up to date. The depth of evaluation will also be influenced by the degree of effort that was applied in the pre-FS and, in this regard, parts of the FS will involve updating and closer investigation of the work performed in the previous studies.

The scope of the FS will depend on the factors associated with a given situation and project characteristics. It typically:

— Evaluates the integration of the NPP into the electric power system in detail;
— Determines the size and main features of the NPP;
— Determines the preferred site and identifies any specific problems associated with the selected site (this might be a separate study or part of the FS);
— Determines which type (or types) of reactor are the basis of bids;
— Carries out detailed cost and economic evaluations and compares them with alternative options;
— Determines the organizational and human resources requirements for implementing the project and operating the plant;
— Determines the overall project schedule;
— Determines the financial viability of the project and the possible sources for financing;
— Determines the contractual approach to be adopted for the acquisition of the plant;
— Analyses the international market for NPPs, fuel cycle and essential materials and services;
— Defines the country's infrastructure requirements and surveys the national participation possibilities;
— Defines the nuclear safety criteria to be applied.

The FS provides detailed analysis and information on all these aspects, with specific recommendations to enable the authorities concerned to make appropriate decisions for the implementation of the project. It also outlines further steps to be taken and identifies areas in which more detailed investigations are still needed.

Finally, the FS provides guidance for the owner requirements for the negotiation of the main contracts or insights for the preparation of the invitation to bid through the development of the BIS.

3.2. APPLICABLE LAWS, CODES, STANDARDS AND GUIDES

3.2.1. Legal and regulatory framework

Initiating and implementing a national nuclear power programme requires a sound legal and regulatory framework covering all matters related to the use of nuclear energy in the country. The government needs to formally commit to the use of nuclear energy solely for peaceful purposes, adhere to relevant international legal instruments, enforce environmental protection, and codify policies for occupational health and the safety of workers and for emergency preparedness and response. The policies will also cover the control of nuclear material, the nuclear fuel cycle and radioactive waste management issues, and nuclear liability. The role of various government bodies, organizations, stakeholders and the public, and their interrelationships in implementing these policies, need to be defined in legislative documents. Implementation of the nuclear power programme will require a regulatory framework for nuclear safety [7], site evaluation, design, construction, commissioning, operation, shutdown and decommissioning of facilities (or closure in the case of disposal facilities for radioactive waste) and radioactive waste management.

During Phase 1 of the Milestones approach, the NEPIO reviews the existing legal and regulatory framework and identifies necessary amendments or additions. This review will include the status of the country's international obligations. The NEPIO incorporates the results of this preview in the pre-FS and the comprehensive report. During Phase 2, a comprehensive nuclear law is enacted, or existing nuclear law is amended to include legal provisions necessary to support the introduction of nuclear power in the country. This comprehensive nuclear law establishes or designates an independent nuclear regulatory body. In this phase, the country also becomes party to all relevant international legal instruments.

The nuclear regulatory body then, in accordance with the legal framework, develops the regulatory framework, including the necessary regulations and guides, in line with nuclear power programme implementation. The regulatory body defines a set of requirements covering all aspects of using nuclear energy through the issuance of regulations. The regulatory body also issues regulatory guides to allow the owner/operator to understand and implement regulatory requirements and to make inputs to the BIS or to the owner requirements document. Before the end of Phase 2, the regulatory body needs to develop the necessary regulations which will be applied during the siting studies and used for the preparation of the bidding documents [1]. These regulations include those defining the licensing process for NPPs covering different stages (i.e. siting, design, construction, commissioning, operations and decommissioning) and associated regulatory guides providing guidance for the implementation of regulatory requirements.

As stated in IAEA Safety Standards Series No. GSR Part 1 (Rev. 1), Governmental, Legal and Regulatory Framework for Safety [7], "The government or the regulatory body shall establish, within the legal framework, processes for establishing or adopting, promoting and amending regulations and guides." In line with GSR Part 1 (Rev. 1) [7], Ref. [1] states that:

> "In setting its requirements, the regulatory body may adopt IAEA safety standards as a reference and may complement these with a well established set of requirements that are in use in States with extensive experience of NPP operation, in particular the vendor country. If the regulatory body decides on this complementary option, the entire set of standards should be carefully reviewed to avoid conflicts, inconsistencies or incompleteness."

In the FS, the owner/operator performs a thorough review of the applicable legal and regulatory framework to identify necessary information to be included in the bidding documents with a commitment to ensure future compliance. The owner/operator eventually communicates with the regulatory body to ensure full comprehension of the current and future legal and regulatory requirements, as this will help to ensure the success of the project. The FS may also provide suggestions to make amendments to the current legal and regulatory framework.

A country's regulatory framework may prescribe applicable codes and standards or leave this to the owner/operator to decide. These include international nuclear and non-nuclear codes and standards. Identification of the applicable codes and standards is an important part of the owner/operator activities during Phase 2, as these are important in estimating infrastructure improvements and programme implementation costs. These may cover the equipment supply and services that are required, and the industrial infrastructure that operates under rigorous quality management programme requirements.

Identifying the applicable codes and standards in the FS helps to assess the acceptable technological solutions, including technical feasibility and cost estimations. Whenever certain assumptions are made (i.e. about certain codes and standards that might be selected as applicable at a later stage), such assumptions need to be clearly identified and justified in the FS.

Depending on the circumstances, specific local non-nuclear regulations, codes and standards may also be applicable to the NPP project. Examples include the local building code; local fire protection requirements; occupational health and safety; electrical design and installation requirements; pressure vessels and pressure piping directives; environmental protection; local land use; drinking and sanitary water regulations; security requirements; foreign investment; taxation; and financial considerations. The owner/operator needs to include a clear description of these regulations, codes and standards in the FS, as they will be necessary inputs for the bid invitation specifications or for the owner/operator requirements document.

Overall, the FS includes all applicable laws and regulations and the list of codes and standards to be applied in the NPP project. Although the compilation of all applicable laws, regulations, codes and standards may be seen as a major task at the FS development stage, this is an effort that also provides the input to the BIS development or to the owner requirements document and will prove to be very beneficial for all parties.

The owner/operator needs to follow the development of the regulations and guides and be involved in the process, as necessary.

3.2.2. Licensing and authorization

The regulatory body established by the comprehensive nuclear law early in Phase 2 normally establishes a licensing process covering the various steps of the implementation of an NPP project, including siting and site evaluation, design, construction, commissioning, operations, decommissioning and release from regulatory control. When developing the licensing process, the regulatory body may make use of relevant IAEA guidance, such as IAEA Safety Standards Series No. SSG-12, Licensing Process for Nuclear Installations [8].

The regulatory body needs to describe the licensing process and associated requirements in all phases of the process (i.e. siting, design, construction, commissioning, operation and decommissioning). The description includes the type of authorization or licence required and all applicable regulations, licensing bases, codes and standards at each step of the process. A timeline for the development of the regulations is prepared by the regulatory body in a graded approach, with priority given to regulations covering the initial project activities up to the construction stage.

The regulatory body also needs to define a typical schedule for the licensing of the first NPP that indicates specific licensing and authorization milestones, supported by the format and content of the technical documentation and specific analysis required for each stage. By defining the format and content of the main application documents, the regulatory body can help to facilitate the owner/operator's development of the documentation to facilitate the licensing process.

An evaluation of the nuclear technologies able to comply with the regulatory requirements needs to be included in the FS for the new NPP. This will allow a preassessment of the NPP designs to facilitate the future licensing process and obtain further insights for the BIS or the owner requirements document.

The regulatory body may decide to sign cooperation agreements with the regulatory body of the vendor country or other countries to help to establish the licensing basis, as well as the required authorizations for siting, design, construction, commissioning and operation, and to define the licensing milestones for each authorization and any related technical documentation.

The owner/operator is responsible for applying for and receiving the licences for the different stages during the implementation of the nuclear power project. Therefore, the owner/operator organization needs to clearly understand its obligations during the licensing and authorization process, to identify all the documents necessary to license the plant, and to plan the sequence of required steps properly (i.e. site permit or authorization, construction licence and operation licence).

The owner/operator's plan for licensing and authorization activities needs to account for other permissions and authorization regimes that are applicable to the design, construction and operation of the NPP. This plan needs to include activities related to compliance with conditions imposed in previously implemented licensing or authorization steps (i.e. site selection or environmental impact assessment approvals, etc.). These also include non-nuclear licences and authorization.

The owner/operator needs to ensure that the potential vendors are aware of the format and content of the various design, safety analysis and other application documentation. The BIS or owner requirements document needs to include specifics on the licensing requirements, along with the obligations of the potential vendor to provide documentation and to support the owner/operator through the process until all the required permits and licences are granted.

3.3. ELECTRICAL SYSTEM ANALYSIS

3.3.1. Review of Phase 1 electrical system analysis

A national or regional cross-boundary electrical system analysis is usually performed during the national electrical system expansion planning study or in a high level pre-FS conducted before the NPP specific FS is undertaken.

The primary objective of the electrical system analysis is to provide a sound technoeconomic background and framework for assessing the electricity production requirements to ensure an adequate electricity supply for the entire time horizon considered. The analysis considers financial constraints, resource availability, and governmental policies and regulations that are currently in place or expected to be in place in the future. It makes projections for the expected demand for electricity and identifies the requirements for new generating capacity to satisfy this demand. Furthermore, it ascertains the timing of investment disbursement and the appropriate mix of generating technologies to supply the electricity required.

In particular, as stated in Ref. [1], Phase 1 studies address the following issues:

"— The capabilities of the existing grid in relation to the NPP technology being considered, including its ability to reliably take an NPP's base load output, its ability to withstand a loss of the plant's output and its ability to reliably supply off-site power during outages and in an emergency;

— The anticipated future growth of grid capacity;

....

— The historical reliability of the electrical grid;

....

— The potential for local or regional interconnections to improve grid characteristics."

During Phase 2, the owner/operator arranges for the available electricity system analysis to be reviewed and updated as part of the FS development to confirm the assumptions made in comparison to the project specific objectives [1]. It is expected that the input data for this update will be more detailed than the data used during the earlier national study of the system expansion plan, including with respect to:

— Current electricity demand and electricity demand projections;
— Current electricity supply, anticipated additions or retirements and future generation capacity expansions;
— Electricity market structure and organization;
— Current transmission and distribution infrastructure and projected changes;
— Evaluation of the impact of nuclear power on the grid.

3.3.2. Updating the projected electricity demand

Long term projections (25–30 years) for electricity demand, peak demand and load shapes are critical determinants of future generation capacity requirements and constitute the reference for any analysis regarding the scope and composition of the electricity supply expansion programme. These include the current electricity demand as well as a review of past trends and projections. The objective of electricity demand analysis is to obtain a basis on which to project future trends, bearing in mind the long term planning horizon and expected structural changes to load patterns, to justify the assumptions made in the NPP project planning.

Many methods have been developed to estimate electricity demand, typically using econometric and/or engineering approaches.

The econometric approach uses statistical methods and historical data to infer the response of electricity consumers to price variations and changes in electricity demand in response to changes in income levels, demographics and other aggregate variables. This approach can capture the behavioural patterns of electricity consumers but is limited by data availability. Moreover, since econometric models are based on historical data, they may not be appropriate to forecast the effects of new technologies or of radical changes in behaviour.

In contrast, the engineering approach emphasizes end use categories of energy demand in analysing electrical load patterns and assessing the impact on load management, efficiency standards and regulatory changes that significantly and directly influence electricity consumption. The focus of this approach on end use scenarios, rather than empirical relationships based on historical data or long time series data, makes it preferable in the case of developing countries with limited historical data.

To assist Member States, the IAEA has developed a special end use model known as the Model for Analysis of Energy Demand (MAED) [9].

3.3.3. Updating the electricity supply system survey

An update of the survey of the electrical supply system is necessary to support the decision to introduce a nuclear power programme into the supply mix. The study in this area covers the current and projected state of the electricity supply systems, the basic characteristics and parameters of the current power plants and transmission grid, and any other foreseen expansions of generation capacity (under construction, committed or planned).

The data obtained from the power system survey, including the technical and economic details, age, fuel type and life expectancy of the existing electrical system (both generation and transmission), are highly reliable, as there are no estimates involved.

The survey also includes an analysis of any past system expansion, in particular regarding the implementation schedules, costs, system availability and load factors, identifying any deviation between the original plans and outcomes along with the justification.

3.3.4. Updating assumptions regarding the electricity market

The electricity market review needs to encompass the following:

— A history of reforms that have led to the current market structure, including market deregulation, if applicable;
— A description of current and historical market participants and commercial arrangements;
— Tariff structure and electricity price components;
— The legal and regulatory environment in which electricity companies would build and operate power plants;
— Barriers to competition;
— Any efficiency promotion initiatives and renewable energy subsidies;
— Environmental issues or regulations;
— Climate change policies and international commitments;
— The current financing and investment environment.

3.3.5. Updating the electricity system expansion plans

Given electricity demand projections and the current electricity supply, a review of electricity system expansion plans is completed to confirm the initial assumptions made regarding the integration of one or more nuclear generating units into the system. The impact of intermittent renewable generation, hybrid systems, energy storage, distributed generation, smart grid concepts and electricity export/import assumptions is also included in the analysis to the extent that these factors affect the project integration requirements.

Special attention is paid to the capability of the transmission system to reliably take an NPP's baseload output, its ability to withstand loss of the plant's output and its ability to reliably supply off-site power during outages and in an emergency [1]. For that purpose, the grid operator, in cooperation with the owner/operator, undertakes detailed studies to determine any expansion, upgrade or improvement necessary to accommodate the anticipated size, technology and site of the new plant. Section 3.4 elaborates further on the studies related to Phase 2 electricity system expansion planning.

3.3.6. Presentation of the results

Presenting results from the updated studies in the FS to stakeholders and decision makers is a critical step, as it involves translating complex technical and economic information into understandable options with their associated consequences. The presentation of the updated electrical system analysis in the FS may aim also to reconfirm the high level conclusions made during the initial Phase 1 technoeconomic analysis.

The owner/operator ensures that the key project outcomes, reconfirmed by updated analysis in the FS, are duly accounted for as necessary in the definition of the project assumptions, the requirements, the implementation time and the funding arrangements. It is also important to demonstrate that the requirements of the planned NPP have been agreed upon with the transmission system operator and that they are compatible with the capabilities of the NPP designs being considered.

The results of the electrical system analysis also identify and provide for planning of the implementation of the enhancements needed to:

— Cope with the enhanced generating capacity;
— Achieve grid stability and reliability requirements to allow safe and efficient operation of the NPP.

3.4. UNIT CAPACITY AND SYSTEM INTEGRATION

3.4.1. Station and unit capacity

The FS includes a section justifying the selection of the nuclear unit size from among those offered by the market. The size of a nuclear unit in this context refers to the maximum electrical power that it can deliver to the grid. The number and size of units selected for the power station influence important factors such as output flexibility and transmission system operations. The number and size of units have to match the current and future expectations for electricity supply and demand as reconfirmed by the electrical system analysis (see Section 3.3).

Considering that the NPP will be integrated into the national electrical power grid, the selection of station capacity considers the necessity of strengthening electrical connections to other nodal points of the electrical grid and to neighbouring countries, and the necessary legal and commercial agreements with those countries. These arrangements and other preparation work need to be completed before the NPP is ready for commissioning [10].

A first nuclear station to be built today within a new nuclear power programme will likely be the one of the larger, if not the largest, single generator in the system to which it is connected. This has implications for power system operations. While economies of scale may lower the cost per kilowatt of nuclear power, the stability of the grid may be vulnerable to disturbances in the output of a single large generator. The addition of one or more nuclear units prompts market operators to preserve the safety and security of the power system in the event of an unplanned shutdown of that generating station. The IAEA models and guidance described in SSG-12 [9] and Ref. [11] may be used in this regard. When small modular reactors are implemented, reduced incremental capacity additions may lower the requirements for grid integration.

The study to determine the appropriate capacity of a nuclear unit ensures that a trip of the unit will not cause a loss of off-site power and that the voltage and frequency of the transmission system remain within the acceptable range. Maintaining system voltage and frequency will ensure continued electricity to the NPP for heat removal and restarting systems after a trip. In terms of impacts on electricity users, a trip of the NPP should not lead to load shedding or increase the risk of system collapse or blackout.

Power systems that add high capacity units may need to procure reserve capacity to replace the capacity of the largest unit in the system should it trip off-line. The amount of available and in-development capacity will also play a role in NPP capacity selection. The planning and operation of a reliable grid are evaluated in detail in Ref. [10].

3.4.2. Integration into the grid

Successful integration of a nuclear station into the grid depends mainly on the size and capacity of the transmission infrastructure. To comply with the conditions to be demonstrated by the grid infrastructure, the owner/operator arranges for an analysis of NPP integration into the grid system, taking into consideration the foreseen plant location and the size and main characteristics of the plant and unit [12]. The analysis takes place in Phase 2 and identifies transmission infrastructure enhancements needed to:

— Cope with the planned generating capacity;
— Ensure grid stability and reliability requirements to allow safe and efficient operation of the NPP.

It is also expected that the grid connection requirements for the planned NPP will be agreed with the transmission system operator and are compatible with the capability of the NPP designs being considered.

3.4.2.1. Grid capacity and grid connection

The siting studies, conducted as part of the site selection process, evaluate the possible candidate sites also with respect to their grid connectivity. Suitable high voltage power lines are required to transmit power from the NPP throughout the grid network, so station size is an important factor when considering transmission capacity. After selecting the plant location, the grid is assessed, not only in terms of capacity, but also in terms of stability. To ensure both, it may be necessary to carry out grid extensions. If the size of the proposed nuclear unit exceeds 10% of the country's minimum electricity demand, the FS will show how demand will be met when the NPP is shut down. It will also outline a plan for system operations capable of ensuring that the grid frequency and voltage will remain within acceptable limits, even in the worst case of a sudden nuclear unit trip.

In addition, the potential nuclear technologies' design safety requirements regarding the availability and reliability of the necessary off-site power supply, such as those concerning capacity and diversity, security and reliability, are studied to ensure that the associated grid enhancement is properly considered and planned.

As a result of the above studies, it is expected that the FS will include information about the planned grid enhancements and arrangements to ensure their implementation in a time compatible with the foreseen construction, testing and commissioning of the NPP project. The planning also needs to include measures to ensure grid stability and frequency control after an NPP trip and for grid emergency situations.

3.4.2.2. Load following limitations

Currently commercially available nuclear power units offer load following capability; however, their technical specifications may limit the magnitude, rate or allowable number of load cycles. Flexibility can be achieved for a nuclear generating unit by varying output, either by a set number of hours, in response to a control signal from the grid operator, or by operating in frequency control mode, in which the output responds to rapid changes in transmission system frequency.

It is expected that during the FS development, sufficient detailed information will be available about the potential NPP designs and respective load following capabilities and limitations. The owner/operator arranges for the transmission system operator to review the planned operating modes and associated load following capabilities. As a result, the operating requirements, including the load following requirements, are specified in the FS as part of the grid related plant characteristics and requirements, and as a basis for their future inclusion in the BIS or owner requirements document.

3.5. SITE AND SUPPORTING FACILITIES

As described in Ref. [1], siting activities begin early in the implementation of a new nuclear power programme. In Ref. [13], the siting process for a nuclear installation is divided into two stages: the site survey in Phase 1 and the site selection at the beginning of Phase 2. In Phase 1, these activities are the responsibility of the NEPIO. The management of siting activities is outlined in Fig. 2 and further elaborated on in Ref. [13].

Early in Phase 2, the regulatory requirements for the site selection and site suitability evaluations are established by the regulatory body. If those requirements are not in place early in Phase 2, credible proxy requirements to use until specific national requirements are enacted are established by the regulatory body and communicated [13]. The owner/operator provides for the screening and ranking analysis so that a preferred site, or sites, can be selected and the site evaluation stage can proceed. This requires selection of the technologies that are to be considered, or at least selection of a range of potential technologies, so that bounding values for any technology specific factors can be considered.

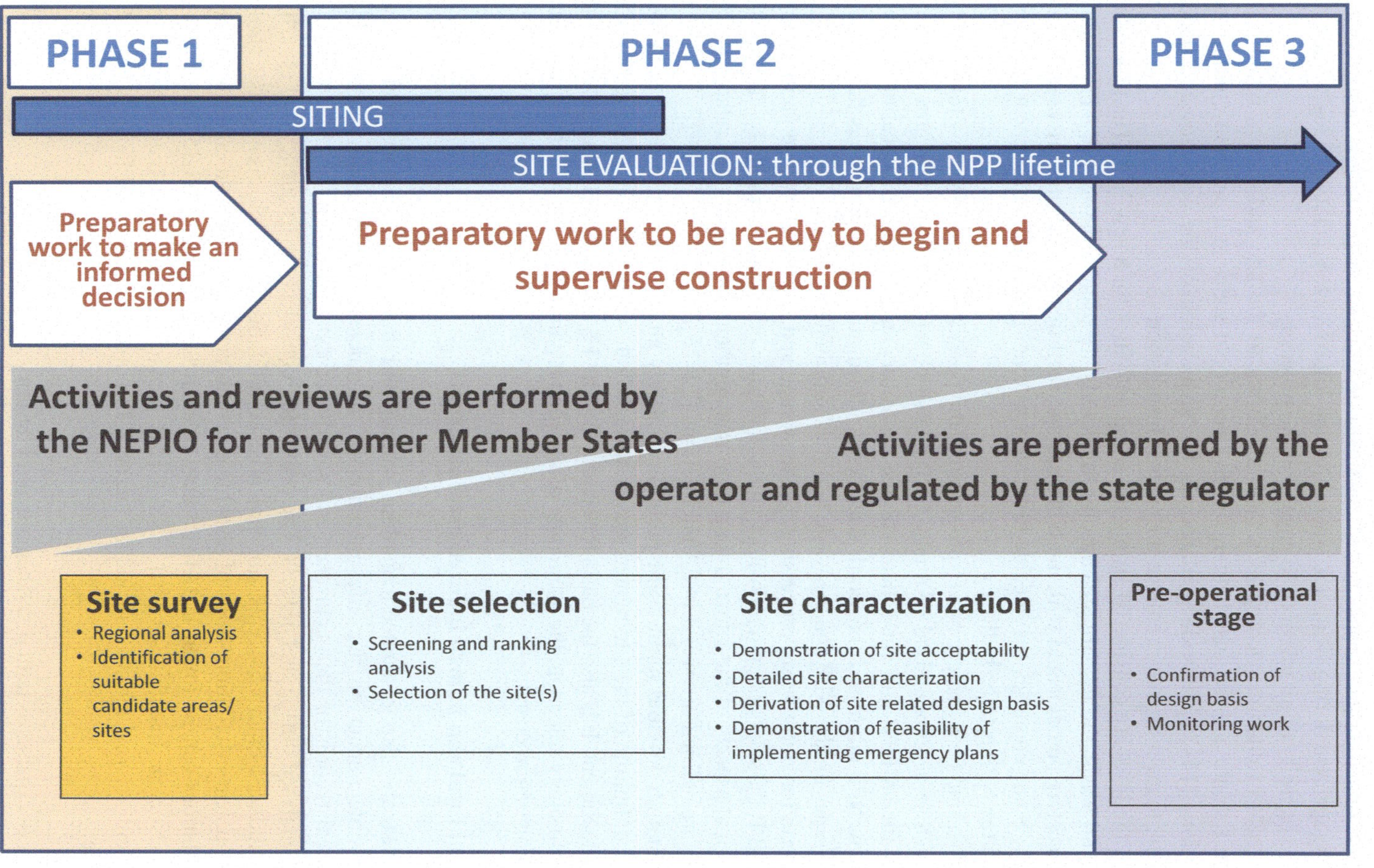

FIG. 2. Organization of the siting and site evaluation processes in the three phases of the Milestones approach. Adapted from Ref. [13].

Much of the site evaluation work then needs to be carried out during the rest of Phase 2, so that the site(s) can be characterized, and the site related design basis is reflected in the BIS or owner requirements document.

As stated in Ref. [13], stakeholder involvement is a necessary and desirable part of the siting process; namely, consulting with and including interested and affected parties in the selection and decision process. Effective stakeholder engagement during siting and site evaluation improves the quality and legitimacy of a decision and builds the capacity of all involved to engage in the process.

The scope of the site evaluation encompasses factors relating to the site and factors relating to the interaction between the site and the installation, for all operational states and accident conditions. Several IAEA publications [14–21] provide guidance on the sequencing of the activities and make recommendations on how to meet the requirements for nuclear installation sites.

Waste predisposal facilities are considered during the siting activities for a new NPP. The processes for the treatment of liquid waste and the storage capacity for liquids are usually part of the NPP design. Such facilities could therefore be colocated with the NPP. Storage facilities at NPP sites are essential, at least for interim storage. Some countries may decide to locate waste or spent fuel management facilities within the site on which the NPP is to be built.

For a country embarking on a nuclear power programme, the possibility of siting a low level waste disposal facility in the close neighbourhood of a new NPP is also considered. It will be cost effective if siting investigations for an NPP and for a near surface repository are combined. Locating a deep repository close to a new NPP might also be worth considering during the siting project, as this option helps to manage siting issues for the NPP and the waste facilities simultaneously.

Although the criteria for siting waste processing, waste storage and spent fuel storage facilities are similar to those for an NPP, the selection of a repository site will require other considerations, with special emphasis on the geological aspects of the site and long term management. Considerations related to the siting of waste facilities are defined in several IAEA publications [22–26].

Detailed site evaluation provides input to the preliminary safety analysis report, if required at this stage, and results in the definition of the site related design basis, which will later be included in the BIS or owner requirements document. It is also the basis for conducting the FS analysis, including defining the requirements for the technology selection, the estimation of the site preparation works and the assessment of the construction and operation costs.

3.5.1. Evaluation of natural and human induced external events

The proposed sites are also evaluated regarding natural and human induced external events to determine their possible impact on the safe operation of the nuclear facility. Relevant available information and historical records of the occurrence and severity of external events, together with site data, are collected and analysed during this process. The evaluation considers the frequency and severity of natural and human induced external events, and potential combinations of such events, that could affect the safety of the nuclear installation. A list of possible external hazards to be considered for the site safety of nuclear installations can be found in IAEA Safety Standards Series No. SSG-35, Site Survey and Site Selection for Nuclear Installations [19]. SSR-1 [27] provides criteria and requirements for the evaluation of different external hazards, including the need to consider changes of hazards and site characteristics with time:

— Seismic hazards (i.e. capable of fault displacement and vibratory ground motion);
— Volcanic hazards;
— Meteorological hazards (i.e. extreme meteorological hazards, rare meteorological events);
— Hydrological hazards (i.e. flooding, low water intake level, blockage of water intake);
— Geotechnical and geological hazards (i.e. landslides, subsidence, soil liquefaction);
— Other natural hazards (i.e. sandstorms, forest fires, combinations of events, etc.);
— Human induced events [14].

Appropriate methods, supported by numerical models, when necessary, are used to characterize the hazards relevant for the site evaluation and design of the nuclear installation. The results of the evaluation of hazards are expressed in terms that can be used as an input for deriving the site specific design basis parameters.

3.5.2. Evaluation of cooling system requirements

SSR-1 [27] states:

"The evaluation of site specific natural and human induced external hazards for nuclear installations that require an ultimate heat sink shall consider hazards that could affect the availability and reliability of the ultimate heat sink."

In addition to the ultimate heat sink safety evaluation, an assessment of the plant cooling system is conducted as part of the FS development activities, since the selected technology can have a major impact on both the economics and the environmental impact of the plant. In the case of water cooled systems, the study includes a sufficient description of the cooling water technology to evaluate water use and its impact on the territory. For plants on a river site, minimum/maximum flows, selection of intake location/invert level with respect to sedimentation/erosion, and the stability of banks are important considerations. For plants on coasts, flood levels/recedes, pumping head, bathymetry, water chemistry and the stability of the shore are important considerations.

3.5.3. Potential effects of the nuclear installation on people and the environment

Specific regional and site characteristics, including the population distribution in the region, are considered, with special attention paid to the transport and accumulation of radionuclides in the biosphere. The IAEA publication on site evaluation [27] provides criteria and requirements for the different elements of the evaluations, including:

— Dispersion of radioactive material;
— Population distribution and public exposure;
— Uses of land and water in the region.

3.5.3.1.   Atmospheric dispersion of radioactive material

Analytical and numerical models and modelling software can be used to perform calculations and assess the atmospheric dispersion of radioactive material releases. Local meteorological and topographical descriptions of the site area, both before construction and during operation of the nuclear facility, are evaluated. The possible impact of meteorological data on the plant design and on its operation — and, conversely, the impact of the NPP operation under normal and accident conditions on local meteorological parameters — are also evaluated.

3.5.3.2.   Dispersion of radioactive material through surface water, including seas and oceans

A description of the surface hydrological characteristics of the region is provided, which includes the main characteristics of water bodies, both natural and artificial, the major arrangements for water control, and information on water intake structures and water use in the region. The site evaluation describes surface water hydrology, including delineation of the drainage basins and available historic hydrological data [20, 21, 25]. An assessment of the potential impact of contamination of surface water on the population is also conducted.

3.5.3.3. Dispersion of radioactive material through groundwater

The ability of the groundwater environment to delay, disperse, dilute or concentrate liquid effluents, as they relate to current and future water users, is evaluated. A conservative analysis of a postulated accidental liquid release of effluents to the groundwater system at the site is conducted to evaluate all potential radionuclide transportation paths with all possible contamination to natural water bodies [20, 21, 26].

3.5.3.4. Ambient radioactivity

Baseline characterization of ambient radioactivity is completed before commissioning of the nuclear facility. It includes baseline data for the atmosphere, hydrosphere, lithosphere and biota in the region, as well as sampling for tritium and other appropriate radionuclides in selected groundwater and seawater systems. The information will be used to allow environmental impact predictions and to provide a reference point zero for monitoring the impact of the plant on the environment.

3.5.4. Population distribution

In this study, the population distribution within the region is described, including current and future projections. The most recent census data and their extrapolations into the future are used as a starting point for the evaluation. The current and future projections of transient residents as valid contributions to the total population counts are considered. The radial sectors for data collection are selected in accordance with the regulatory radial distance guidance.

A low population zone starting from the centre point of the proposed facility is projected and shown on a map on which topographic features can be added, such as roads, railways and waterways that may be used for evacuation purposes, together with the locations of all public facilities and institutions. Special attention is paid to vulnerable groups, as this is a key indicator of successful emergency planning. The population within the low population zone is provided to include estimates of daily and seasonal peaks of permanent and transient populations.

With regard to emergency planning, SSR-1 [27] states:

"The feasibility of planning effective emergency response actions on the site and in the external zone shall be evaluated, with account taken of the characteristics of the site and the external zone as well as any external events that could hinder the establishment of complete emergency arrangements prior to operation."

3.5.5. Use of land and water in the region

For the purpose of preparing emergency plans, the domestic use of land and water is evaluated to assess the potential effects of the nuclear facility on the region. Dominant land uses and metropolitan centres surrounding the sites are reviewed and evaluated from the viewpoint of water usage safety. The investigation also includes land and water uses that may serve as habitats for plants and living organisms in the food chains of humans and animals [20, 28–30].

3.5.6. Preliminary site layout and site preparation considerations

The IAEA Milestones approach [1] expects that within the site studies performed during Phase 2, necessary improvements for local infrastructure at the preferred site or sites — such as access, services and facilities — will be identified, and plans for implementation will be developed. This includes evaluation of the necessary arrangements, costs, implementation schedules and sequences that need to be considered in the development of the economic assessments of the FS. Some specific elements to be considered are elaborated in more detail in this section.

3.5.6.1. Preliminary site layout

The conceptual layout for the various temporary buildings and facilities required during the NPP construction phase is established. These include the concrete batch plant, the soil reception area for filling and backfilling, the field manufacturing facility, the bonded and general warehouses, and administrative offices. Several alternative layouts are also conceived, especially for structures presenting the possibility of several solutions — namely, preliminary layouts of the intake and outfall canals, the radioactive waste building, the switchyard and others.

For the case where the NPP is to be constructed in a coastal area, breakwater defences for protecting the nuclear facility from abnormal sea waves are considered. The requirement for a harbour or wharf is established, and the selection of an optimal location for heavy equipment loading and unloading is evaluated. This can influence the NPP's general arrangement and layout alternatives.

3.5.6.2. Constructability and logistics considerations

At the preliminary site layout stage, an evaluation of the construction process and the construction method is conducted. The study includes accessibility requirements for the assigned module fabrication area; the ringer crane to move heavy equipment; a laydown area; the crane type, its location and interference with adjacent facilities; and a site access strategy with entry, exit and check points for equipment and personnel at each building.

The layout features the shortest possible distance between the equipment delivery wharf and the power plant entry point. The heavy equipment transportation route for components such as steam generators, the turbine generator stator, the nuclear reactor, the main transformer, the emergency diesel generator and condenser modules are reviewed, and safety is checked for possible soft interferences and crossovers of sensitive underground utilities.

3.5.7. Summary of site characteristics and supporting facilities

In summary, the FS report:

— Confirms the conditions for the acceptability of the preferred candidate site(s) (i.e. site licence/ approval may not be required at this stage) in terms of minimizing all safety hazards associated with natural phenomena and human induced events and in terms of the absence of potential negative effects of the nuclear facility on the environment, and in consideration of local population characteristics and the facilitation of emergency planning;
— Describes all physical characteristics of the site(s) related to the safety of the NPPs, including, but not limited to, seismology, meteorology, geology and hydrology;
— Identifies the design basis for natural and human induced external events (e.g. design basis earthquakes, design basis floods, design basis meteorological events, aircraft crashes, explosions, pressure waves, fires);
— Addresses the radiological impact on the population and on the territory during normal and accident conditions, including dispersion characteristics in air and water and the feasibility and effectiveness of the necessary emergency preparedness and response plans;
— Describes the exact site location, a preliminary layout of the NPP, administrative facilities, worker camp facilities, property boundaries, etc.;
— Establishes and describes the plant cooling system, including its preliminary layout and the grid integration plan, and describes the available infrastructure, such as roads, ports and airports.

SSR-1 [27] provides detailed information on this topic.

3.6. ENVIRONMENTAL IMPACTS OF THE PROJECT

According to Ref. [28], one of the basic principles underpinning the beneficial, responsible and sustainable use of nuclear energy is that people and the environment are protected in compliance with the IAEA safety standards and other internationally recognized standards. In this regard, the approach to environmental protection in new nuclear power programmes is discussed in Ref. [29].

3.6.1. Environmental impact assessments and development of the feasibility study

Environmental impact assessments (EIAs) are usually standalone documents prepared separately from an FS. The FS authors thus consider the main EIA findings regarding environmental protection to evaluate costs and risks. Normally a project specific EIA is performed before the FS. In this case, the EIA report is included as a reference in the FS report, with the EIA conclusions and recommendations elaborated in the FS. It is important to include sufficient information to justify the EIA's assumptions and requirements that have implications for the project execution, schedule, construction, and operation and maintenance (O&M) costs.

If an EIA is not available at the time the FS is being prepared, a succinct assessment of the environmental impact is conducted within the FS. The environmental assessment section in this case includes:

— An introduction;
— A listing of the regulatory requirements;
— A description of the environment characteristics and the proposed plant;
— A projection of the environmental impact of the plant during construction, operations and any design basis, as well as the severe accident condition;
— The proposed management and monitoring plan for the environment;
— The ultimate consequences of the project for the environment and the overall conclusions.

3.6.2. Framing environmental protection within the nuclear infrastructure development process

As described in Ref. [30], and shown in Fig. 3, the environmental management process consists of certain steps in each phase of the development of the nuclear infrastructure.

In Phase 2, the three main steps lead to the preparation of the following documents:

(1) The environmental scoping report (ESR);
(2) The EIA;
(3) The environmental management plan (EMP).

Each of these steps is described briefly below. More information on this topic is provided in Ref. [30].

The first step in Phase 2 is for the owner/operator of the proposed NPP to organize the development of the ESR, which will form the basis for the execution of the EIA. The ESR builds on the initial environmental information collected in Phase 1, or from a strategic environmental assessment, if available [29], and identifies the requirements for additional information and analysis to complete a comprehensive project specific EIA. The ESR aims to identify all aspects for which there may be impacts that will be assessed in the EIA. The ESR also identifies key stakeholders and provides a stakeholder participation and communication plan. Depending on national requirements, the ESR may be shared for review by stakeholders, including the public, and may be submitted for approval to the competent authorities.

When the ESR is finalized, the owner/operator prepares, or arranges for the preparation of, the EIA report. Using the ESR as the basis, baseline environmental information is collected to form the starting point for assessment of the impacts caused by the project. A complete EIA report includes consideration

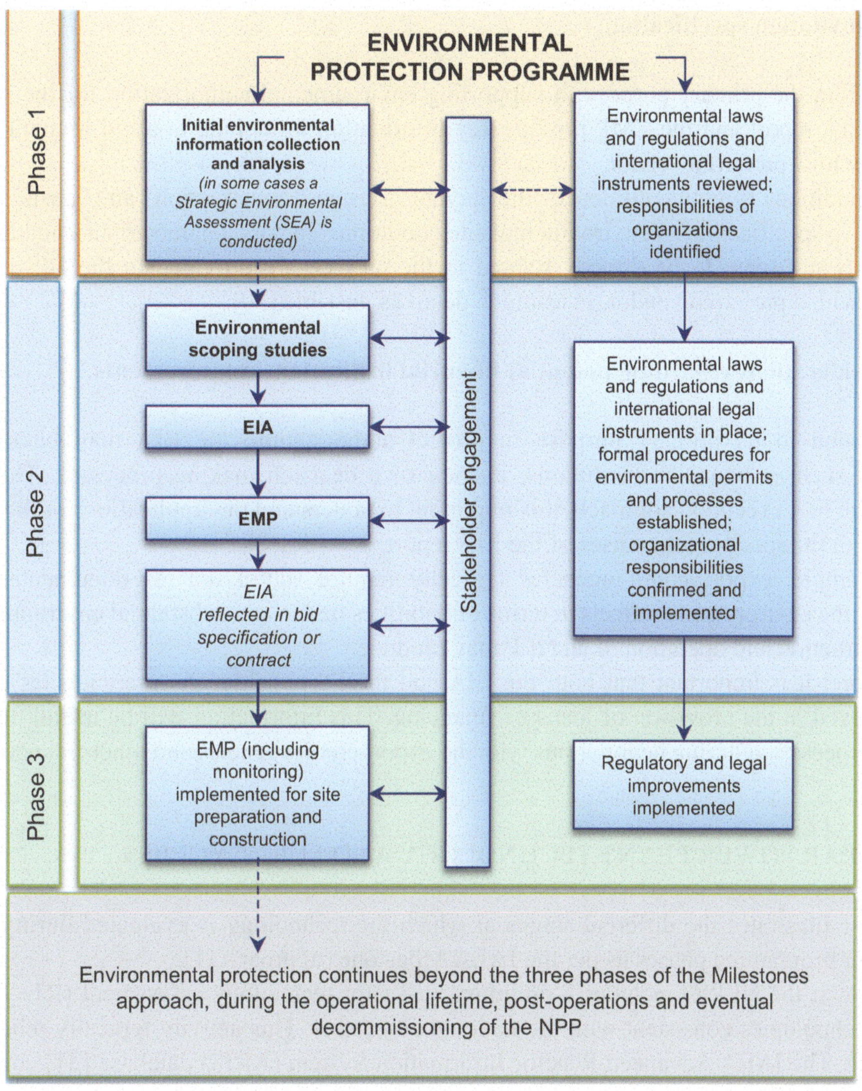

FIG. 3. Steps in the environmental management programme according to the Milestones approach.

of both nuclear and non-nuclear impacts. The environmental and nuclear regulatory functions in many countries are the responsibility of separate bodies. In such cases, the nuclear regulatory body plays a role in the evaluation of the nuclear part of the EIA in coordination with the environmental regulatory body and other relevant governmental agencies.

An EMP that incorporates mitigation measures identified during the EIA process is then prepared. The EMP forms both the guideline and a legally binding document addressing environmental issues throughout the project life cycle from pre-construction through operation.

The draft EIA report including the EMP is customarily submitted for review by stakeholders and to the competent authority (or authorities). During the public consultation process, adjustments and amendments to the EIA report may be requested by the stakeholders and required by the competent authority. Final approval by the competent authority represents acceptance of the analyses and conclusions in the EIA report regarding the environmental impacts and management measures for the NPP project.

To address the issue of uncertainty in the final design of the plant, including the likelihood that the vendor is unknown at the time of the EIA report preparation, the plant parameter envelope concept may be used. The plant parameter envelope uses enveloping values for all technologies under consideration for those aspects identified to lead to potential environmental impacts.

3.6.3. Bid invitation specification

Apart from the primary purpose in supporting environmental authorization for the proposed NPP project, the EIA report and the EMP provide key information to include in a bid invitation or to guide negotiations with a preferred vendor.

Site conditions greatly influence the layout, design, construction and costs of the NPP. Comprehensive specification of environmental site conditions, factors, characteristics and data, including those that may not seem to be directly related to the project, are provided in the BIS or included in negotiations with a preferred vendor, in as much detail as possible [30].

3.6.4. Considerations regarding potential financial institutions' requirements

In addition to its primary purpose in project authorization, the EIA may be used by some stakeholders, such as financial institutions, in their own decision making processes. Therefore, when developing the EIA execution approach, it is important to understand the implications of the requirements associated with the specific future uses of the EIA report.

For example, export credit agencies typically require submission of documentation to allow evaluation of the environmental effects in terms of liabilities tied to any substantial environmental damage that the construction and operation of an NPP may produce.

Therefore, it is important that both the EIA and the FS consider the prerequisites of institutions that are involved in the provision of loans or financing. This information will be useful during the loan application process and during negotiations with the export credit agencies and other financial institutions.

3.7. NUCLEAR POWER PLANT TECHNOLOGY AND FUEL CYCLES

Figure 1 illustrates the different stages at which the technology is evaluated during the different nuclear power programme phases as per the IAEA Milestones approach [1].

In Phase 1, the NEPIO includes a preliminary reactor technology assessment (RTA) in its pre-FS to identify technologies consistent with the national objectives. This activity typically relies on publicly available data. The IAEA Advanced Reactor Information System (ARIS) database [31] compiles reactor technology options in a structured way to facilitate the preliminary RTA exercises.

During Phase 2, the owner/operator:

— Revisits the preliminary assessment performed in Phase 1 by the NEPIO.
— Engages with the vendors to collect more detailed information on their designs and their relevance for the host country. This privileged information may be obtained through a request for information (RFI) process, as mentioned in Ref. [2].
— Conducts a more comprehensive RTA based on newly collected information. The IAEA has developed a detailed technology assessment methodology for this purpose [11].

The three steps above provide input for the section on nuclear technology and the fuel cycle section of the FS report. Section 3.7.6 details the expected final content of the section.

3.7.1. Input and organization

To assess reactor technologies and to develop its own criteria, the owner/operator needs input from the national strategy developed by the NEPIO in Phase 1, in particular:

— The target year for commercial operation;
— The grid capacity at this horizon;

— The expected nuclear legal and regulatory framework regarding radiation protection, safety, security, safeguards, emergency preparedness and response, transport, radioactive waste and spent fuel management, and export and import control provisions;
— The applicable legislation regarding environmental protection, occupational health and safety, and foreign investment;
— The site survey with the siting possibilities already screened;
— Constraints on technology options resulting from the considered financing options;
— The level of local industrial involvement;
— The considered options for the national fuel cycle strategy.

These elements are expected to be found in the pre-FS, along with other studies and assessments in Phase 2. Preparation of this part of the FS is an iterative process in which the owner/operator provides elements to stakeholders such as the NEPIO, the regulatory body or the grid operator as it assesses technologies for near term deployment. The need for such interfaces is described in Ref. [3]. Similarly, the RTA section in the FS report has an influence on the other sections.

The owner/operator sets up an initial RTA team in charge of this section of the report and defines the interaction that they will have with the other contributing staff to ensure consistency in the report. This RTA team will perform the market survey and gather elements from vendors and elsewhere. The RTA team will become knowledgeable about the relevant technologies, will develop a BIS or an owner requirements document and will eventually discharge the role of construction oversight during Phase 3. This evolution of the organization is further described in section 5.2 of Ref. [3].

3.7.2. Nuclear power technology market survey and technology assessment

It is advisable for the owner/operator to include a market and industry survey in the technology assessment section of the FS report. After collecting the input as described in Section 3.7.1, the RTA team reviews the preliminary RTA performed by the NEPIO to identify the range of reactor designs that can fulfil the policy objectives established during Phase 1. This work will require the owner/operator to collect additional information on available designs.

Therefore, it is advisable for the owner/operator to launch an RFI process to allow vendors to share privileged information on their design and experience, including project deployment assistance. An example of an RFI is provided in appendix I of Ref. [2]. Another benefit of initiating an RFI is that information can be requested in a structured manner and thus be compared more easily afterwards.

The owner/operator can then perform an assessment of the information gathered and include it in the FS report. Reference [11] describes the process of and steps for technology evaluation to be carried out by the owner/operator.

Typically, the NPP technologies are explored sufficiently in the FS to demonstrate their financial and technical values and their comparative benefits, as the FS is intended to identify all conditions necessary to meet the programme policy objectives and the NPP project goals. This will be helpful in the subsequent establishment of the acceptable ranges of design characteristics and other requirements to be specified in the BIS.

If a separate RTA document is not produced, a sufficiently rigorous assessment is still included in the FS because it is necessary to describe, categorize and select the preferred technologies before negotiations start, either through a tender or through direct negotiations.

Beyond compliance with the local regulator's safety requirements, the IAEA safety standards, safeguards requirements and physical protection provisions, the evaluation included in the FS report also demonstrates:

— What design features of each technology meet policy objectives and project goals;
— How the constraints from policy objectives and project goals beyond the vendor proposed design features can be accommodated, either by activating design options or by more structural changes.

3.7.3. Fuel supply and irradiation cycle parameters

For each reactor design under consideration in the FS, the corresponding fuel cycle implications are covered. The nuclear fuel cycle broadly covers fuel supply, reactor operations and radioactive waste management. Policy objectives and project goals related to the fuel cycle need to be identified, and their specific economic, safety and environmental impacts assessed.

According to the IAEA Milestones approach [1], it is expected that during Phase 1 the NEPIO will have identified preliminary approaches that may be feasible for the country regarding the nuclear fuel cycle, covering all stages from the front to the back end and taking account of the economic, safety, security and non-proliferation implications.

During Phase 2, the owner/operator defines a realistic front end nuclear fuel cycle strategy over the short and the long term, based on the national policy and Phase 1 work.

Accordingly, the FS refers — at a level of detail that is appropriate for assessing the impact on project feasibility — to such matters as the number of initial core reloads to be requested from the vendor in the NPP contract, and the procurement strategy for the fuel services (i.e. natural uranium, conversion, enrichment and fuel manufacturing) needed thereafter, together with planning for construction of any intended front end fuel cycle facilities consistent with the national long term fuel cycle strategy, the power plant construction project and the national non-proliferation commitment.

For each reactor technology or specific NPP design under consideration, the corresponding nuclear fuel irradiation cycle implications are identified, including the main parameters such as annual fuel requirement, power operation cycle length and discharge burnup [11]. All these parameters are input data for the FS and affect the plant and refuelling operations, the plant economics considering the length of the refuelling cycle and the reactor downtime, and the handling and storage capacity requirements for spent nuclear fuel.

3.7.4. Spent fuel management

Spent fuel management comprises the different steps required to ensure that spent fuel is safely managed after it is discharged from the reactor core. It includes a period for initial cooling, subsequent interim storage, transport, reprocessing and recycling (if this is the selected spent fuel management strategy), conditioning and disposal of the final waste form, which may be spent fuel final waste canisters and/or the high level radioactive waste form resulting from spent fuel reprocessing operations.

The selection of spent fuel storage option(s) is a critical step at the back end of the fuel cycle, and it is a decision that needs to be made in relation to the subsequent steps of the spent fuel management strategy. Spent fuel storage arrangements depend on several physical features (e.g. initial enrichment, burnup, residual heat, maintainability) and non-technical factors such as the national spent fuel management strategy, stakeholder involvement, international obligations and economic considerations [32–35].

During Phase 1, the NEPIO analyses possible options for spent fuel and radioactive waste management so that decisions taken at the end of Phase 1 are well informed about the challenges that spent fuel and waste create for a nuclear power programme [1].

Based on the national policy developed in Phase 2, the owner/operator defines the requirements related to the implementation of the national back end fuel cycle strategy that affects the plant operation, including the facilities for storage (on-site and off-site) that are constructed as part of the project, possible reprocessing or arrangements for fuel take back, the strategy for interim spent fuel storage, transport and ultimate disposal [12].

3.7.5. Management of radioactive waste

For the purpose of the FS, developing a good understanding of the immediate and long term responsibilities that arise as part of generating nuclear power is an important prerequisite. Consideration is also given to ensuring provision for exemption and clearance of material from regulatory control.

According to the IAEA Milestones approach [1], at the onset of Phase 2, the NEPIO or the government establishes a radioactive waste management policy and strategy.

It is expected that during Phase 2 the owner/operator will develop a strategy based on this national policy and convey the associated requirements for radioactive waste processing, storage and disposal (including spent fuel if considered as waste), as well as the decommissioning plan, to potential vendors.

Solid, liquid and gaseous radioactive waste will be generated continuously throughout the operating lifetime of the NPP and during its decommissioning [35]. All types of waste need identified processing, storage and disposal schemes that take into account the characteristics of the waste as it is generated and over the duration of any storage period.

Solid waste comprising process and laboratory waste (e.g. personal protective equipment, wipes, filters, contaminated equipment, spent ion exchange resins) are typically sorted, segregated and, if needed, size reduced and then packaged and stored pending disposal. Liquid waste (e.g. cooling water, storage pond water, secondary waste coming from decontamination) is usually segregated, interim stored and then conditioned into a stable matrix suitable for disposal. Gaseous waste (e.g. tritium, krypton) is typically discharged to the environment in a controlled manner according to strict licence conditions.

Measures are established and implemented to avoid and reduce the volume of radioactive waste that is produced at all life cycle stages of NPP operation and decommissioning.

Although decommissioning of a new NPP will start many decades after commissioning, it is prudent to have a decommissioning strategy from the outset that considers the types and volumes of waste that will arise and will need to be disposed of. Therefore, the FS typically will have a section discussing the requirements for development of the decommissioning strategy as part of the NPP design package and a preliminary estimation of the associated costs.

Providing for the future liabilities of decommissioning and radioactive waste disposal requires the establishment of a funding mechanism, typically through a tariff placed on the price of electricity production. A significant part of these long term liabilities reflects the costs of implementing the various disposal programmes needed for operational and decommissioning low level waste, as well as for the intermediate level waste and high level waste. Methods to estimate such long term disposal costs and the associated funding schemes are presented in Ref. [36].

3.7.6. Technology recommendations

Based on the work performed for the technology, fuel cycle and waste management assessments, this section of the FS report:

— Provides a summary description of the work that has been performed in the assessment of the NPP and fuel cycle technology (description of the process and outcomes).
— Identifies the key parameters affecting the results of the assessment and details all key elements from the programme policy objectives and the NPP project goals, along with their priorities and justification for methods used in the evaluation and ranking of reactor technologies.
— Presents an assessment of the NPP designs available in the market, including their implications in terms of fuel and generated waste (strategies and funding and technical requirements), against the conditions set in the pre-FS.
— Provides a justified recommendation on the designs to be considered in the selection process, aligned with the national strategy defined in Phase 1 and consistent with the other parts of the FS. This includes applicable laws and regulations, grid capacity and characteristics, site and heat sink characteristics, power economics, training requirements, national industrial involvement and project implementation.
— Presents an overview, based on the above mentioned findings, of the preparation of the BIS in the case of tender or of the owner requirements document in the case of exclusive negotiation (e.g. under intergovernmental agreement).

To ensure the overall consistency of the FS, the characteristics of the recommended reactor designs, along with the range of values, assumptions and conditions to secure nuclear fuel supply, spent fuel management, radioactive waste management and disposal, are passed on to the teams in charge of other evaluations within the FS.

3.8. NUCLEAR COST ESTIMATE

The costs of an NPP project span over several decades and extend well beyond the operational lifetime of the plant. It is important that all these costs are correctly accounted for in the lifetime cost of an NPP project. In the FS, the objective of the owner/operator is to obtain estimates of the various costs to the extent needed based on the available information.

Cost estimation is the process of developing an approximation of the financial resources needed to cover the lifetime project cost. As discussed in Ref. [37], cost estimators take inputs from the project's preliminary scope, schedule, cost structure, risks and external factors to produce an estimate for the various project activities and the project as a whole. Experience from previous similar projects can also provide valuable inputs. To complete this part of the FS, the owner/operator needs specific knowledge and expertise, which typically is arranged with the support of external specialized organizations possessing proven experience.

The International Energy Agency (IEA) and the Nuclear Energy Agency (NEA) have developed a systematic and comprehensive framework to account for the different costs occurring over the lifetime of an NPP [38]. As shown in Fig. 4, the framework breaks down the lifetime costs of an NPP into three main cost components:

(a)   Capital investment costs, which are incurred during the preconstruction and construction phases and occur before the NPP generates any electricity (and thus any revenue);
(b)   Nuclear fuel cycle costs;
(c)   O&M costs.

Cost categories (b) and (c) occur mostly when the power plant is in operation and generates electricity and revenues. In addition to these costs, countries having NPPs typically require the operators to contribute to a dedicated fund or funds covering the costs of waste management and decommissioning after the shutdown of the plant. In this process, the account system serves as a checklist to verify and benchmark the estimation results.

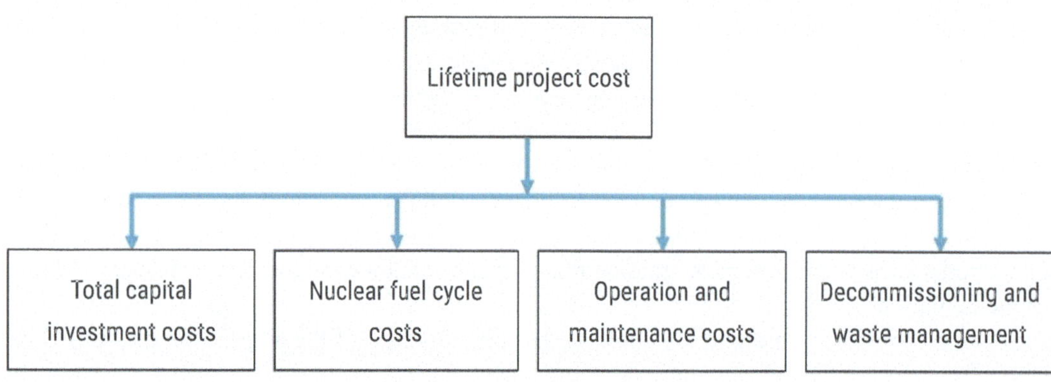

FIG. 4. Lifetime project cost breakdown.

3.8.1. Total capital investment costs

Total capital investment costs (TCICs) are the costs of acquiring and building an NPP and bringing it to commercial operation. These costs constitute most of the total lifetime cost of electricity from nuclear power (roughly 75% of the total costs for a discount rate of 7%). The TCICs can be broken down into two broad categories: overnight costs — the cost of building an NPP if it were literally constructed 'overnight' (no financing and no escalation); and financial and escalation costs. This breakdown is important, as failure to distinguish between estimates that include financing costs versus those excluding them can lead to confusion in the comparison of published capital cost figures.

In the IAEA/NEA account system, overnight costs comprise accounts 21–54 and 70 [39]. The overnight costs concept is useful because it abstracts from the reality that the construction of an NPP will typically stretch over several years, during which the price of equipment, labour and materials may well escalate and interest charges will be incurred on borrowing to finance construction. A cost estimate that does not make such an abstraction will, by definition, embed forecasts of escalation, construction duration, expenditure profiles and possibly interest rate variation. Since these variations are likely to differ between projects, cost estimates that include them make it difficult to compare project costs in a useful manner. Differences in overnight costs between projects can be related to differences in labour costs at various sites, material requirements for the construction of different types of plant, etc.

Reported overnight investment costs vary widely, even when presented on a per kW(e) basis. A 2020 IEA/NEA study [38] reports overnight costs for nuclear plants for 'nth of a kind' projects in Organization for Economic Cooperation and Development (OECD) and non-OECD member countries. Other reference data are available in earlier surveys [40–43].

The project specific overnight costs assessment in the FS may be based on either a top down or a bottom up cost estimation approach. The former approach relies on the application of escalation indices and engineering scaling factors to a 'reference plant'. A reference plant is a plant that is technologically similar to the proposed NPP, but which was constructed at a different time and place (and thus under different market conditions), even though the reference plant may differ in some significant respects. Escalation indices and engineering scaling factors would be applied to adjust for differences in relevant market conditions, plant size, etc.

A bottom up approach to estimating potential NPP costs relies more on the decomposition of a plant into its major components. Some of these components (e.g. turbine generators) may be more transparently priced than others, but with expert advice or qualified support it may be possible to gain understanding of the overall costs of the complete NPP construction project using this approach. The owner/operator also makes sure that, irrespective of the approach to estimating overnight costs, the costs for project development and project management are included, as well as any other owner's costs, such as those for site acquisition, site and infrastructure development, licensing and permitting. Reference [36] elaborates on cost estimation approaches and specifics.

The financial and escalation cost component of TCICs can be broken down into escalation costs (account 60), interest during construction (IDC) (account 61) and fees (account 62). Escalation costs result from changes in the price of commodities, wages, etc., the estimation of which can be a challenging task. Similarly, IDC — which comprises the accumulated money disbursed to pay interest on borrowed capital — is forecast in any cost estimate prepared for the purposes of evaluating project feasibility from an economic perspective. Both escalation costs and IDC will depend, to a large degree, on the duration of construction.

For new builds in embarking countries, the least risky scenario is that of selecting a plant for which there is significant construction experience. In this case, the cost of the new build can start from the detailed cost of the reference plant that can be known with a fair degree of accuracy (assuming that such data are available). National regulations and licensing practices vary from one country to another and are to be considered when establishing the cost of the new build. The estimators will have to add the cost of the inevitable detailed design changes necessary to adapt the reference design to the new site conditions and the cost of any new licensing conditions. In addition, any profit margins and price escalations for the equipment, construction and incidentals will have to be considered.

3.8.2. Nuclear fuel cycle costs

The IEA/NEA accounts cover the major drivers of fuel costs for both once through and reprocessing fuel cycles, including front end uranium supply, conversion, enrichment and fuel assembly fabrication. Accounts are also defined for the costs of interim storage and final disposal of the spent fuel in the case where no reprocessing takes place, as well as the costs of interim storage, reprocessing and final disposal of radioactive waste in the case that reprocessing is used at the back end of the cycle. In estimating the fuel costs likely to arise over the lifetime of an NPP for the purposes of economic evaluation during an FS, it will be necessary to assume which of these two fuel cycles will be employed.

Fuel costs are included in items 100–171 in the IEA/NEA account system. Estimates of the total fuel cost, including front end and back end components, once through and plutonium recycling fuel strategies, are given in the Refs [40–43].

3.8.3. Operation and maintenance costs

O&M costs comprise costs such as wages and salaries for NPP staff (engineers, technical support, O&M and administrative personnel), as well as consumables, materials, refurbishment and repair, purchased services, insurance and taxes. O&M costs comprise those incurred in the management and disposal of low and medium (intermediate) level radioactive operating waste. O&M costs are included in items 800–890 of the IEA/NEA account system.

Current estimates of O&M costs vary significantly depending on the country considered and the type and size of the reactor considered, as well as whether a single unit or multiple units are considered. Estimates presented in the referred surveys [38, 40–43] also show the evolution of these costs over time.

3.9. PROJECT IMPLEMENTATION APPROACH

The project implementation approach is an important section of the FS and contains the analysis of different management areas that need to be assessed for a proper NPP feasibility analysis. It also identifies areas of strength for the project implementation, and eventually areas of weakness that need improvement before the completion of Phase 2. As described below, the FS needs to present information on the management system with an adequate description of each management area. This information may be included directly in the FS report or by references to other documents that the owner/operator has developed and implemented, such as the quality management system, project planning and scheduling systems, and contracting approaches for the new NPP.

A sound and effective management system at the level of the owner/operator is an essential element for a proper feasibility assessment. In cases where the management system is not fully implemented when the FS is elaborated, plans are made to address gaps and shortfalls. References [37, 44] present useful and comprehensive information on project implementation and organizational structure, which are reviewed during the FS for the relevant inputs and references.

3.9.1. Ownership structure

The FS considers the types of ownership structure that will support the feasibility of the NPP project, depending on the country specifics. The preferred ownership structure is described to help inform the decision at the end of Phase 2 in relation to the new NPP implementation. The following examples of ownership and commercial structures are discussed in Refs [45, 46]:

— The sovereign model, under which the new NPP is developed, built and operated by a state controlled entity.
— The corporate based model, under which the new NPP is developed, built and operated by a corporate entity operating on (market based) commercial principles. Such a corporation could be controlled by the public sector or have shareholders from the private sector.
— The project based model, where the new NPP is developed, built, financed and operated by a special purpose vehicle, which has been established for that purpose.

Variations in the above three models are possible, providing flexibility for potential innovative techniques that are crafted to suit a specific case. The economic and legal/regulatory environment can put a range of constraints on these models.

In the selection of the ownership structure, the following issues will have to be considered:

— Possible preference for ownership models based on majority shares for the founding company;
— Identification of possible financial and strategic partners;
— Possible cooperation in the economic, financial, technical and procurement areas;
— Risks associated with each ownership structure and their management.

Generally, embarking countries use a centralized ownership structure based on the governmental atomic energy commission, or a nuclear utility created for the purpose of developing the NPP and managing operation as a single owner. One approach could be that of assigning the ownership and management of the nuclear new build to a project company responsible for the construction and operation of the new NPP. At a later stage, the project management company may consider selling a percentage of its shares to strategic and/or financial investors. These investors could provide expertise, funding and electricity off-take. Another approach would be to allow large energy consumers looking to hedge their energy bills and secure lower electricity costs in the future to take partial ownership. The relationship between the different models and project risk allocation is discussed in Ref. [46].

3.9.2. Contracting strategy

The selection of the contracting strategy is one of the basic decisions to be taken concerning the realization of the NPP construction project. It therefore receives great attention and is based on a careful analysis of all aspects of the FS and/or technology assessment, as well as other factors, such as owner/operator experience. The IAEA provides guidance and reference information on the different contractual approaches applied to NPP construction.

Three contractual approaches are discussed in Ref. [2]: the turnkey approach, the split package approach and the multi-contract approach. In addition, the IAEA Nuclear Contracting Toolkit [47] provides useful details, templates and other references. Further guidance is provided in Ref. [44] on how to evaluate and define the contractual approach for construction of the NPP, considering that the contract type can greatly influence the performance goals, the implementation schedule, the supply chain (including local participation) and the cost estimates for the project. Factors include:

— Potential vendors and their experiences and attributes;
— Standardization and proven quality;
— Government and industrial relationships;
— Competitive and economic considerations;
— Foreign financing possibilities;
— Guarantee and liability considerations;

— Planning and implementation of the project and subsequent projects;
— Availability of a qualified project management, coordinating and engineering workforce;
— Development of national engineering and industry capabilities;
— Owner/operator experience in handling large projects.

Generally, for an embarking country, the turnkey or EPC contract approach is preferred. This approach contributes to the positive feasibility for building the plant, as it assigns full responsibility for all engineering and construction activities to the EPC contractor, relieving the owner/operator of the management of the project and coordination of the interfaces between major suppliers. A turnkey or EPC contract typically includes the complete scope of work required to construct and commission the plant. The EPC contractor is responsible for designing, constructing and commissioning all elements of the operating NPP, and for ensuring that the plant complies with the owner/operator requirements, permitting and licensing requirements and meets all specified performance requirements, such as guaranteed capacity and efficiency.

Nevertheless, in a turnkey or EPC contract approach the owner/operator remains responsible for certain elements. It is important that the contractual documents specify a description of the scope of work and the allocation of responsibilities, including the following:

— Preparing, reviewing and adapting the necessary project planning and implementation schedules;
— Establishing the overall requirements, monitoring progress and approving the main engineering tasks;
— Preparing local infrastructure, such as an emergency preparedness plan and an information centre;
— Preparing a public education and awareness programme on operations, incidents, accidents, radiation exposure, etc.;
— Maintaining effective project cost control;
— Carrying out plant design reviews to ensure adherence to contractual conditions and regulatory requirements;
— Introducing and coordinating quality assurance programmes (integrated management systems);
— Reviewing the quality assurance programme for all contractor(s);
— Ensuring quality control and proper construction supervision at the plant site;
— Conducting surveillance of component manufacturing;
— Interfacing with regulatory authorities and applying for plant licences and revisions;
— Reviewing and approving plant safety and engineering procedures, as well as plant O&M manuals;
— Implementing the plant security plan;
— Preparing for commissioning and operation;
— Supervising plant commissioning and reviewing test results.

3.9.3. Procurement programmes

Procurement of the plant and services related to its construction, such as siting and consultancy services, are key owner/operator activities during Phase 2. In developing the procurement programme, the owner/operator considers the following:

— The selected contractual approach, as discussed above;
— Requirements, practices and experience related to procurement in the country;
— All laws, rules and regulations that govern interactions in the procurement process.

As elaborated in Ref. [44], at the end of these preparatory activities upon approval of the FS results, the owner/operator is capable of starting a bidding process or direct negotiations in the case of a preselected technology provider.

The owner/operator procurement programme is aligned with the procurement techniques applied to NPP new builds and involves oversight of the main contractor's activities, including verification of project progress and quality requirements. In consideration of the selected procurement approach (open tender, competitive dialogue or direct negotiation), the owner/operator procurement organization usually establishes several documents to guide procurement. These may include the procurement criteria and planning, supplier qualification and selection, execution of the tendering and bid evaluation process, contract award, contract and claims management, and other necessary procedures. If the contract involves a split package or multipackage procurement approach, a significantly greater level of competence and effort will be required on the part of the owner/operator.

The FS evaluation assesses the owner/operator's plans for acquiring the necessary resources and capabilities. These plans may include the appointment of an owner's engineer to support the owner/operator organization. All these elements of the owner/operator procurement programmes and planning are presented in the FS, together with an assessment of the associated human and organizational resources, as well as time and cost requirements.

3.9.4. Project schedule

The construction schedule is an important element for the project feasibility analysis that is included in the FS. The schedule proposed by vendors may be optimistic and significantly shorter than those historically achieved. Therefore, it is important to assess the adequacy of the schedule estimates by determining:

— If the key assumptions are valid and if all critical assumptions are represented;
— If the schedule scope includes all pertinent activities;
— If task durations are realistic relative to historical precedent and current standards;
— If the schedule logic is sequenced in a reasonable manner;
— If modularization is used in the design and whether its consequences have been incorporated;
— If the critical path scope is complete and the logic is reasonable;
— If the vendors have performed a risk assessment and what the significant conclusions were;
— If human resources have been considered in the different scheduled activities, as this is a key element to judge the feasibility of the schedule for performance and achievement.

In the pre-FSs during Phase 1, there is usually not much information on plant engineering and construction schedules, and therefore the NEPIO relies on international benchmarking experience in NPP construction. During Phase 2, the owner/operator becomes familiar with the principles of planning and scheduling, as discussed in Refs [37, 44], for a sound analysis of the schedule for the construction and commissioning of the NPP.

At this stage, it is expected that the owner/operator may use preliminary schedules from vendors who have expressed interest in bidding as a basis, and will assess their feasibility from the point of view of the assumptions that pertain to the specific conditions under which the project will be implemented. As it is difficult to compare schedules across different technologies and vendors, the feasibility assessment focuses on completeness, adequate level of detail, critical paths and key assumptions that meet the project strategic target, and reviews the following factors:

— Status of the domestic infrastructure to support the schedule;
— Elements supporting the overall schedule improvements compared with previous experience;
— Impact of project management and organizational structures;
— Influence of modularization and construction planning;
— Interactions with regulators and new regulatory processes;
— Impacts of key vendor assumptions;

— Effects of plant staffing on construction and operations;
— Incomplete design and engineering issues.

Incomplete information regarding design suitability, outstanding technology issues and insufficiently defined scope areas may have a negative impact on the reliability of project schedule estimates. Significant schedule risks are inherent in projects involving the initial units of a first of a kind design and/or designs in which vendors have introduced significant changes. Even conceptual designs that have received certification or approval from their country of origin regulators may still be lacking complete design details, which can affect the timely delivery of components. An important consideration for FS evaluations is that nth of a kind designs with substantial track records reduce uncertainties in the schedule.

3.9.5. Project management

3.9.5.1. Project management structure

During Phase 1, the NEPIO typically addresses only the top level implementation approach and organizational and project execution structures. For the FS developed in Phase 2, the owner/operator performs a deeper analysis of the implementation approach, weighting advantages and disadvantages, and making a thorough analysis of the owner/operator responsibilities, as discussed in Ref. [3], and the project management tasks, as described in Ref. [37].

Although the owner/operator organization may not be fully established when the FS is initiated, the important issue is to determine the future tasks in all the project management areas and to identify gaps. The establishment of an adequate project management structure by the owner/operator is a prerequisite for achieving Milestone 2.

3.9.5.2. Project management tasks

Project management tasks are described in Refs [37, 44]. The main activities of the owner/operator in Phase 2, which could be a reference for the FS development, are also described in Ref. [3].

The owner/operator's tasks include:

— Preparation of project planning and execution schedules;
— Establishment of the local infrastructure;
— Preparation of the public information programme;
— Preparation of the basic environmental protection data;
— Project progress assessment and management;
— Introduction and implementation of the quality assurance programmes and quality management system;
— Execution of project cost control;
— Supervision of manufacturing;
— Contact with the local authorities;
— Review and approval of the safety and technical procedures, as well as of the O&M manuals of the power plant.

The project management responsibility ends with the handover of the completed and functioning plant systems, structures and components to the organization responsible for the O&M of the completed project.

3.9.6. Engineering and design

In the contractual approaches of turnkey and split contract packages, the engineering and design is performed by the EPC contractors. However, there are activities that require owner/operator capabilities in engineering and design, such as oversight and review of engineering deliverables by the contractors, and oversight of third party engineering companies that are involved in the NPP project. In addition, the owner/operator understands the licensing requirements and establishes efficient communication with the licensing authority during the safety approvals for the new NPP.

A preliminary evaluation of engineering capabilities in the country is performed, and the results are included in the FS report. Examples of areas requiring local engineering participation could be support for the conversion of the nuclear technology vendor detail drawings for civil construction in accordance with the local standards to obtain the civil construction permits. This is also valid for fire protection, and eventually for mechanical (pressure components and systems) and electrical systems, depending on the country's local regulations for these subjects, and/or in compliance with regional requirements that may apply to the country.

Experience has shown that local participation in construction may involve engineering resources that can adapt and apply vendor documents from the country of origin. At the feasibility level, it is important to evaluate whether sufficient capacity exists or if the EPC contractor will make special arrangements affecting the schedule and costs and increasing risks for the project.

The IAEA provides guidance through standards and publications related to engineering and design. A series of important safety design requirements and guides are provided in Refs [48–61]. It is not expected that the owner/operator will be fully proficient in all these IAEA standards and guides at this stage. However, as this is the starting point to become a 'knowledgeable customer' and to create the basis for the future 'design authority' role when the plant starts operation, it is advisable that some assessment of the related in-country capabilities for engineering and design is included in the FS.

The BIS includes the requirements for technology and knowledge transfer from the vendor that will later support the creation of the design authority function in the owner/operator organization during project execution, as per the International Nuclear Safety Advisory Group's INSAG-19, Maintaining the Design Integrity of Nuclear Installations Throughout Their Operating Life [62].

3.9.7. Construction

Irrespective of the contractual approach, the construction of a new NPP relies on substantial human resources to be sourced in the country in which the NPP is being built, and possibly from neighbouring countries or other sources. Therefore, the feasibility analysis for construction needs to consider the availability of the necessary resources and their capability to perform the activities to stringent specifications and quality requirements.

In this regard, the guidance on construction in Refs [37, 44] regarding health, safety and the environment, lessons learned, security and safeguards are considered during the FS development. The guidance on the application of innovative construction methods in Ref. [63] may be complemented by the guidance in Ref. [64]. IAEA Safety Standards Series No. SSG-38, Construction for Nuclear Installations [65], provides advice on introducing a safety culture and a graded approach during construction, and identifies organizational and technical aspects that differentiate NPP construction from conventional industries and that may be considered in the FS.

3.9.8. Commissioning

Although commissioning is not usually a 'project deal breaker' at the feasibility level, the FS includes an evaluation of the commissioning specifics, considering that it is a project stage that ends up with plant turnover to the operating organization and that requires the deep involvement of the owner/operator from the management and technical standpoints, including consideration of commissioning readiness in the resource and training plans.

In the case of turnkey contracts or split package contracts, the contractors train and develop the owner/operator personnel on plant O&M. The operating personnel need to be actively engaged during the construction and commissioning activities to gain practical experience to complement their theoretical knowledge of plant structures, systems and components. The commissioning resource planning starts when the main contracts are signed and the allocation of personnel is in accordance with the project construction schedule.

The IAEA has established useful standards and guidance related to commissioning, which may be considered within the project implementation planning and as inputs for the bid invitation specifications and background for negotiation with contractors [66–68].

3.9.9. Quality assurance and quality control

The need for a sound quality management system for new NPP construction has been addressed by many IAEA standards and guidance: Ref. [44] has a section dedicated to quality planning and management, Ref. [37] also has a section dedicated to quality management and Ref. [63] has a section discussing quality assurance, inspections and testing. The establishment of a strong safety culture and quality management by all parties involved in the NPP construction is highlighted in the project management section of IAEA Safety Standards Series No. SSG-38, Construction for Nuclear Installations [65].

During the FS preparation, the owner/operator conducts a deep assessment and evaluation of the quality management systems and capabilities of the supply chain. The assessment examines the adequacy of the quality management system of the owner/operator and the availability of related quality procedures among potential vendors and contractors. In cases where the quality management system is not yet fully established, the FS includes conclusions on its implementation status, and the corresponding plans and projections, as well as their adequacy and the availability of the necessary resources. Reference [69] provides information for developing and assessing the quality management system for nuclear installations, and this information is considered in the FS preparation by including insights as necessary.

3.10. PROJECT RISK MANAGEMENT

NPP construction projects are major projects with long schedules and large budgets. Their complex nature makes them prone to various risks that may incur cost or schedule overruns or even derail the project. Thus, there is a need for effective project risk management to mitigate such risks and their operational and financial consequences.

Risk management for new NPP projects is based on a detailed and comprehensive analysis of a broad range of characteristics of the project and associated risks. Reference [70] provides detailed guidance on an integrated risk management approach that includes identifying risks, developing strategies to manage risks and implementing these strategies at the early design and construction stage of a facility while considering its entire life cycle.

During the FS, it is expected that a detailed analysis of the investment project under consideration and an investment plan will be completed by the owner/operator, including a detailed assessment of the project's economics and resilience to risk, which is further elaborated in Section 3.13.

The owner/operator will benefit from adopting a systematic approach to identifying all potential project risks and applying techniques to manage identified risks. Identification of the probability and consequences of various issues or events allows optimal decisions to be made to minimize adverse effects and maximize social and business objectives in a cost efficient manner.

At the FS level, it is important to identify major risks that can affect the project with eventually catastrophic results (i.e. the project to be stopped) or that pose high impacts, such as long delays in the project schedule or budget overruns, contractors that change their line of business, or a licensing framework that is not stable and subject to numerous changes. The FS includes a brief analysis of major project risks with the associated measures for mitigation or neutralization, in such a way that the positive outcomes of the FS are not altered by risks that appear during project execution, as well as the assumed principles of the risk management during future project development stages.

3.10.1. Risk identification

Many different information sources can be used to identify risks, such as industry specific or generic risk exposure checklists, flowcharts of critical processes, examination of contracts, physical inspection, analysis of financial statements, and employee, contractor or regulator interviews. Lessons learned during previous experiences in NPP construction are useful for general risk identification and categorization. A major objective of risk identification is to avoid the unintentional or unconscious retention of risk that occurs when a source of performance variability remains undiscovered and is therefore not part of the risk assessment/management system.

After risks are identified they are characterized and assessed. Some qualitative questions can help the owner/operator examine the essential characteristics of the risk from a conceptual point of view:

— What are the factors that trigger risk?
— Is it possible to intervene in the factors that trigger the risk?
— Does the risk produce both opportunities and threats, or only one of these? If both, do we need to measure both?
— Is the cause of risk likely to be continuously occurring or is it episodic or rare in time and space?
— Is the risk such that a risk management decision/action will be reversible in the future or is it likely that for this source of risk, the choices are basically irreversible?
— What are the potential effects of the risk on the overall FS?
— Is the source of risk such that it is mission critical ('make or break') or is it a source of risk that will modify results in less severe ways?
— What are the financial effects of the risk?

Examples of the risk identification process can be found in Ref. [71].

3.10.2. Techniques or strategies to manage risks

The risks identified and characterized are next evaluated with respect to the best combination of techniques for their management. Three generic categories of risk management techniques are:

— Reduction of risk;
— Retention of risk;
— Transfer of risk.

In practice, one or more of these techniques is likely to be used in managing risks associated with a particular issue. It is also important to examine whether the use of a particular solution takes into account the interaction among different areas of risk. In all cases, as stated in Ref. [37], ultimately the project risk rests with the owner/operator. Therefore, the owner/operator needs to actively manage the various risks inherent in the project and make conscious decisions regarding their management.

3.10.2.1. Reduction of risk

Reduction of the risk involves the identification of measures to reduce the likelihood of the occurrence of the risk and/or measures to reduce its consequences if it does occur. Reduction of severity can include measures to keep events from progressing into more severe episodes, as well as measures to reduce the economic impact of severe disruptions. These risk reduction measures may be pre-event, simultaneous with event, and/or post-event actions.

3.10.2.2. *Retention of risk*

Even though the owner/operator may wish to transfer risks to other parties, some risks are best handled by the owner/operator. These risks are considered when applying the risk retention management technique.

3.10.2.3. *Transfer of risk*

Risk transfer means that the original party exposed to a loss can obtain a substitute party to bear the risk. The idea is to allocate the risk to the party who can control the results, or prevent the problem, or manage the risk if it happens, or can best absorb the impact. Most risk transfer mechanisms consist of some form of contractual agreement with a counter party.

3.10.3. Risk plan implementation techniques

Another aspect of the risk management framework is to plan the implementation of the chosen techniques or strategies. As part of the FS risk evaluation, an assessment of the effectiveness of the chosen approach can be tested by questions such as:

— Does the strategy or solution address the identified risks?
— Is the selected solution consistent with the solutions to other risks?
— Are the key risks addressed by the selected strategy?
— Can the withdrawal plan be exercised?
— Is flexibility maintained?

3.10.4. Risk plan implementation monitoring

The final step of the risk management process is to monitor the results of the risk management plan implementation and to provide feedback (positive or negative) so that risk analysis is always updated as the operating environment changes. The results from the monitoring activities are fed back to the appropriate step to continually improve the risk management process. It is expected that the FS will reflect that the approach to monitoring risk management plan implementation is one of the key elements of project feasibility.

3.11. STAFFING AND TRAINING REQUIREMENTS

This section of the FS includes a study of the plan to acquire and develop the required workforce of the owner/operator organization, securing an adequate number of highly qualified and experienced personnel and maintaining their high level of competence over time. As part of the inputs to the FS, the total estimated investment in human resources will be identified.

To achieve this, this section of the FS addresses influencing factors such as the plant location with respect to population centres and contractor support services, the regulatory requirements for construction and operation, the number of units planned at the site, the environmental monitoring requirements, the effort required to assimilate and implement construction and advanced construction techniques and modularization, future plant staff qualification and training programmes, labour laws, and the effort required for public awareness and education.

3.11.1. Staffing requirements

An effective workforce plan for the owner/operator covers the organization, including policies and structure required to support the construction and operation of the plant. This plan builds on the national human resources recruitment and development plans developed in Phase 1 by the NEPIO. The owner/operator needs to assess both the national plan and its own organizational plan during the FS development activities to ensure that the resources needed for its planned activities will be available. The cumulative resource profile through all project phases from beginning to end typically takes the shape of an 'S' curve. Reference [72] provides estimates of the overall resources for the key organizations during different phases of a nuclear programme.

One of the first activities to be performed by the owner/operator is an analysis of the competencies required to support the activities to be conducted in Phases 2 and 3. Detailed descriptions of the typical functions and activities of the owner/operator during Phases 2 and 3 are presented in Refs [3, 73]. As presented in Ref. [73], the workforce plans will differ in each country that develops a nuclear energy programme, based on factors such as the following:

— Government policies (e.g. staff and materials import restrictions, localization strategy);
— Scope (and main purpose) of the construction programme;
— Nature of the construction build programme (fully turnkey versus transition to indigenous construction) and subsequent relationship with the vendor;
— Availability of nuclear expertise from non-energy applications (e.g. industry, medical, agriculture, applied sciences);
— Access to available international nuclear expertise;
— Existing (if any) nuclear educational programmes;
— Availability and quality of non-nuclear workforce.

The owner/operator designs its planned organization and estimates the number of staff and the competencies required in each position. The owner/operator organization then creates a human resources development plan to acquire the needed competencies to carry out its mandate and meet its goals for safety, security, safeguards, quality and efficiency during each phase. Human resources development programmes include elements of training, on the job experience and exposure to international good practices.

As the owner/operator organization expands from a project development organization in Phase 2 to a construction and commissioning organization in Phase 3, the human resources needs will change in terms of both staffing and competencies. The responsibilities and capabilities of the owner/operator and their evolution throughout the NPP project implementation are presented in Ref. [3]. During the FS development, these principles need to be studied in detail, together with the information collected from potential vendors and reference plants, to make sure that the owner/operator's plans for organization and staffing consider the latest international experience and best practice. Typically, the approach to recruitment and training will follow a phased approach.

While many of the owner/operator's activities are best done in-house, it may be more appropriate for some tasks to be outsourced. The determination of which activities will be performed in-house and which will be outsourced will have an impact on the organization's recruitment plan and its training programme. If the decision is made to perform a given activity in-house, the development of the required competencies may necessitate early consideration of recruitment and training, especially for those positions with long lead times.

To make effective use of outsourcing, the owner/operator needs to retain sufficient in-house competency to act as a knowledgeable customer with the ability to specify the work to be performed, supervise its implementation and evaluate the results. Embarking countries that outsource activities can use this as an opportunity to develop their own competencies by including an element of training (e.g. shadowing, mentoring, on the job training) in the contract.

During Phase 2, the owner/operator organization is focused on establishing and developing the organization itself and carrying out siting activities to support the preparation of documents necessary for the EIA and the site licence application. Also, during Phase 2, the owner/operator needs to develop its understanding of commercially available nuclear power technologies and to establish technical specifications for the NPP to be included in the bid invitation or to guide negotiations with a preferred vendor. At this time, it will also have to define the management system, the safety culture and the communication strategy, and start implementing new arrangements, as identified in Phase 1, for strengthening the infrastructure for emergency preparedness and response.

During Phase 3, the owner/operator manages and oversees the EPC contract and the construction and prepares for operation by recruiting and training the operating staff. The staff requirements in Phase 3 are much larger than those during Phase 2 and require training programmes that can last several years. The typical number and specialities of the owner/operator organization throughout the project implementation phases are presented in Ref. [3].

Actual staffing needs for plant operation depend on many factors, including the size of the plant, the type of technology, the services outsourcing policy, the availability of local support and other local specifics. Average staffing levels of approximately one person per megawatt installed (MWI) for a large single unit plant are provided in Ref. [73]. For small units, the specific staffing level tends to increase to approximately 1.5 persons per MWI, while stations with several large units are expected to reduce the specific staffing level to approximately 0.7 persons per MWI. Actual staffing numbers from operating plants in various countries, as well as projections for future small modular reactors, are provided in Ref. [73].

The initial operating organization can be staffed by vendor trained local staff with vendor expert supervision, by turnkey contractor staff, by local staff from other plants and local trainees, or by a mixture of experienced and newly trained staff in strategic positions. These models usually include a gradual handover schedule.

With respect to the training schedule, the training of the staff used for commissioning would naturally have to be completed before the turnover from construction to commissioning. More than for any other group, they may have to receive hands on training in reference plants abroad. The actual training needs will vary depending on the specific technology, the level of automation, maintenance, layout, national laws and regulations, regulatory requirements and the level of interfacing.

Because the competence requirements of the workforce evolve through the different phases of the life cycle, including operation, and to mitigate the risk of knowledge loss through staff turnover, measures to use construction, commissioning and operation experience and to retain knowledge are put in place.

3.11.2. Training requirements

The duration of training programmes may range from a few weeks of familiarization for already trained personnel who will continue to be employed in their specialized field, to several years for a plant operator or a trainer in radiological safety radiology who may have had no relevant work experience.

Other than nuclear science and technologies, the training programmes include environmental protection, radiation safety and health physics, handling of radiological and hazardous materials, nuclear and industrial waste, legal and regulatory requirements, safety culture and questioning attitudes, event reporting relevant to safety and quality, industrial safety and specialized labour skills. In all organizations (including contractors), an appropriate systematic approach to training model is applied to ensure that the training provided will support development of the necessary competencies.

The FS development needs to assess the resources and facilities required for the training itself and the deployment of nuclear expertise. The resources and facilities will have to be sized to match the scope of the training requirements. For countries embarking on nuclear power for the first time or those expanding their programmes, special attention is paid to the possibility of recruitment either from conventional industries or directly from educational institutions.

In Phase 3, most of the plant specific training will be provided by the vendor through the EPC contract. The development of the training programme needs to consider that some job positions require long lead times and, possibly, language training. It is important to recognize that for many job positions, actual experience at an operating NPP is required in addition to classroom training. This applies not only to control room operators but also to several specialist and managerial positions. Therefore, typically the training plans are developed for each job position in a systematic way by application of the systematic approach to training model [3].

External expertise can be used, either by awarding contracts to experienced consulting organizations with the obligation to train national recruits during work package delivery or by incorporating experienced consultants into the local organizations, working side by side with local staff.

When national recruits are sent abroad to build competencies, this is usually carried out through bilateral agreements with governments, regulatory agencies, vendors or other utilities. Training may take the form of first principles basic training in educational institutions via formal courses or at different levels, such as vocational, undergraduate and postgraduate programmes. The IAEA offers training courses, fellowships and internships. The FS needs to reflect the assumptions on how the required training will be organized and to assess the impact of these assumptions on the project implementation, contracting strategy, schedule and cost estimations.

3.12. NATIONAL PARTICIPATION

The national participation level is one of the major focus areas when developing a nuclear power programme. Many goods and services are required to construct an NPP and to support its operation. Such supporting activities can be a source of jobs and economic growth for the country. They can also help to transfer technology to the country, increasing the technology knowledge base. A sound nuclear industry base will support the NPP during its many years of operation, allowing it to manage and resolve maintenance and engineering issues using in-country resources.

To achieve a high level of national participation, the local industries need to be capable of meeting the strict codes and standards and the rigorous quality programmes that are associated with goods and services destined for an NPP. While maximizing the participation of domestic industrial and manufacturing resources is the common aspiration of many countries embarking on a new NPP project, national involvement in NPP development is a gradual growth process that requires a policy and an action plan with a sufficient budget favouring the evolution of national capabilities.

Encouraging the participation of domestic industries requires States to formulate and establish policies and incentives to build the capacity and capabilities of national industries, along with the needed partnerships to extend local involvement. Considerations to arrange industrial involvement to support the NPP construction programme are discussed in Ref. [74]. For that purpose, several selected activities are performed carefully before the project execution phase starts. They can be carried out as part of the pre-FS or the FS, or in separate specific studies, and their results are included in the FS report. These activities include:

— A detailed national industry survey;
— Establishment of a strategy for national participation;
— Development of a localization plan;
— Development of a technology transfer plan.

As stated in Ref. [74], before Milestone 1 (ready to make a knowledgeable commitment to a nuclear power programme), the NEPIO typically performs the following activities to establish the national policy and goals with respect to national participation:

— Establishes strategy for optimizing local industrial involvement;
— Conducts initial survey of local industrial organizations;

— Initiates dialogue of the owner/operator (if determined) with suppliers, and the government to government consultations;
— Develops a localization plan as part of the pre-FS.

By Milestone 2 ('ready to invite bids/negotiate a contract for the first NPP), the owner/operator or the project owner builds on provisions to implement the established policy, which includes the following actions:

— Expanded survey of local industries, including auditing of their capabilities;
— Continuation of the dialogue with suppliers;
— Determination of localization provisions to be included in the EPC contracts.

In preparing the FS, the owner/operator could refine or reassess the local participation based on investigations of the localization and business appetite. The NPP owner/operator also typically uses the national participation targets and assumptions as an input to the development of the BIS. The NPP owner/operator specifies the requirements for national industrial involvement in the BIS documents, including the plans for the implementation of technology transfer from the vendor [2].

3.12.1. National industry survey

An assessment of the national and local capabilities to supply commodities, components and services for the construction of a nuclear facility provides an essential input to further economic, financial and commercial studies. Since the NPP to be constructed needs to meet high safety requirements, there is a need to identify industries within the country that can satisfy such strict codes and standards or would be willing to upgrade their management and manufacturing systems.

It is best to begin a national level industrial survey with an assessment of the present engineering, construction and industrial capability, as well as participation in conventional power projects and in other types of large project in the country and abroad. The national industry survey would lead to useful conclusions or recommendations that can then be included in the FS report, together with a list of local companies with capabilities to support the first NPP. The list also indicates the level and type of quality management in these companies.

If the capability is lacking in some aspects, the survey includes, as applicable, requirements for upgrades, which may include, but are not limited to, the quality assurance programme, the acquisition of new technology and know-how, the installation of additional equipment, innovative inspection techniques and improvements in methods and procedures, including computer aided design and manufacturing. Any upgrade generally implies extra financial demands on these organizations. Investments, interest and service charges associated with the consequent development effort are evaluated on a cost–benefit basis.

As stated above, during Phase 2 and the FS development, the main responsibility for establishing the local industrial involvement begins to shift from the government to industry. The main organizations that are expected to have the lead roles in this effort include the following [74]:

— The prospective owner/operator of the planned NPP unit(s);
— Local industrial organizations that provide goods and services to the existing electricity generation industry;
— Professional organizations that represent local engineering, construction and manufacturing industries;
— Trade unions whose members work in the above industries;
— Vocational education and training organizations that develop the skills of the local workforce;
— Organizations that provide technical and scientific support, including goods and services, to the existing local nuclear application industries (e.g. non-destructive examination, radiation protection, dosimetry, quality management).

The survey is therefore focused on these industrial organizations and determines their optimum ability to provide the needed services and goods while adhering to strict standards. In conducting the survey, the financial and economic limitations of the industries that provide goods and services also need to be carefully taken into consideration so that the optimum participation of the local industries can be determined.

Local engineering and research companies, including universities and material testing and calibration laboratories, are also explored as potential future TSOs for the construction, commissioning and operation of the first NPP and for the further implementation of the national nuclear power programme. Based on this investigation, a specific upgrading programme for the candidate TSO could be developed, including training and support, possibly by the NPP vendor. The results of the various national industry surveys provide an input to the project cost estimations and to the sections of the FS report discussing local participation and the project implementation approach.

3.12.2. Establishment of the strategy for national participation

An overall localization strategy aimed at encouraging national participation, covering the short to long term, is developed based on the results of the national industry survey described in Section 3.12.1. The strategy may focus initially on conventional items within the capability of local manufacturing entities and define a plan for industrial development and acquisition of know-how to supply nuclear grade items later.

Initial national participation considerations and targets are typically subject to assessment within the FS performed by the owner/operator during Phase 2. At this stage of the project development, more information is available concerning the candidate NPP technologies and the potential contracting and execution strategies. The national and local capabilities to supply commodities, components and services for building and operating an NPP on schedule, at competitive prices and with appropriate quality controls and assurance may be assessed with a high level of confidence.

Development of the strategy needs to consider the following advantages for the country of optimizing national participation:

— Improved industrial competitiveness and self-sufficiency;
— Increased local employment;
— Improved national engineering capability;
— Increased ability to use new technology and acquire know-how;
— Strengthened capability to independently train the local workforce;
— Support from government agencies, ministries and industrial or commercial organizations.

The following steps are to be taken in the development of the strategy:

— Establish clear goals and objectives (e.g. engineering, construction and equipment installation, manufacturing, commissioning, TSOs), based on the state of national industry and service suppliers in the conventional power sector.
— Determine the degree to which local industries with their existing know-how can meet the national participation objectives.
— Identify what new technologies and facilities will be required to achieve the envisaged local participation goals and determine the needs for technology transfer.
— Ensure that the period required to obtain the necessary local capabilities and skills is consistent with the nuclear power programme schedule, always keeping in mind that national participation can never compromise the quality and safety aspects of the plant.

3.12.3. Localization plan

The localization plan represents an action plan within the overall national participation strategy described in Section 3.12.2. It can either be included in this section of the FS report or be part of a separate study, with only the conclusions reported in this section.

During the development of the localization plan, the following important aspects are to be taken into consideration:

— The required development and quality management upgrades of the potential local supplier and any associated costs.
— The difficulty of transferring technical knowledge of complex equipment to organizations where manufacturing know-how is not adequate. This transfer can be accomplished through suitable technical assistance or technology transfer agreements from experienced technology suppliers, accompanied by suitable training.
— The cost of providing support to potential domestic suppliers with the basic capability to manufacture a product but requiring some additional equipment and technical know-how.
— The cost of creating prototypes and the cost of developing industrial capability whenever precision components of nuclear grade have been produced on a repetitive basis.
— The convenience of adopting a standardized design that makes repetitive manufacture possible, thereby progressively cutting costs and time required.

The information included in the localization plan section of the FS represents the assumptions to be used as the basis during negotiations with the NPP vendor on the expected level of involvement for local contractors.

3.12.4. Technology transfer

Technology transfer is a learning process that is only successful when knowledge, skills and experience have been imparted to people in the receiver organization. One of the most important topics to be considered carefully in the FS report is the degree of technology transfer, which refers to the development of indigenous capabilities related to NPP O&M, to component and system design, and the manufacturing of equipment and special materials.

A technology transfer and training programme comprises several steps, and may be agreed between:

— Government agencies and organizations;
— An NPP vendor and NPP owner/operator;
— Technical expert organizations;
— Research institutes (scientific cooperation in energy sectors);
— Universities and technical schools;
— Industries and utilities in the fields of design, construction, component manufacturing, maintenance and operation.

Generally, in a nuclear power programme, there are three basic technologies that can be transferred:

— Design technology: This can start at the R&D stage and continue right up to final process design of all the systems that comprise an NPP.
— Manufacturing and construction technology: This includes the design of nuclear equipment and extends through the special manufacturing techniques and quality assurance methods to the construction techniques for an NPP.

— Project engineering and project management: This deals with the work necessary for the successful execution of a nuclear power project. It includes both home office and activities on-site, and also the pre-project period.

Many embarking countries will find that it is not feasible to develop a highly complex technology in a reasonable time using solely domestic resources. Acquisition from abroad is the usual method for obtaining nuclear technology. Where localization is not yet possible, technology transfer is the first step to be explored. Normally, this requires governmental involvement to secure continuity, typically ensured by a bilateral cooperation agreement sealed in conjunction with commercial contracts for the NPP. The contract itself will usually cover technical and managerial cooperation between the NPP purchaser and the NPP vendor.

The owner/operator can develop the technology transfer plan, either in the context of the FS or in a separate study whose conclusions and recommendations are included in the FS. After the FS has been approved, the information on localization and technology transfer can be incorporated into the BIS so that it may be negotiated in detail with the selected NPP vendor and included in the commercial contract.

Given the potentially sensitive nature of some nuclear technology, special considerations apply to technology transfer from foreign suppliers to domestic manufacturers or engineering companies, including, but not limited to, compliance with international treaties and instruments related to safeguards and export control.

3.13. ECONOMIC AND FINANCIAL ANALYSIS

The funding and financing[3] requirements for an NPP are significant, and financing is a major hurdle for an NPP project, especially in countries lacking previous experience with nuclear power. Financing for a first NPP can be pursued in several ways, such as sovereign (government) financing, corporate balance sheet financing by the owner/operator, export financing from the vendor country, project financing or a combination of these. The key differences among these options are the ownership pattern that they establish, which in turn governs the degree to which they protect the interest of investors and creditors, and the ways in which they allocate risk. Developing a successful plan to obtain such financing will require significant expertise.

In Phase 1 of the programme, specific analysis would have been performed to determine the funding requirements as a function of time for the following elements:

— Initial infrastructure;
— Sociopolitical acceptance of the NPP;
— Creation or hiring of required expertise or consultants;
— Creation and maintenance of a competent regulatory body;
— Creation of expertise for competent project management and operating staff;
— Security and safeguards arrangements;
— Long term storage of radioactive waste;
— NPP decommissioning;
— Human resources development.

An evaluation of financing options for the NPP implementation has been performed in Phase 1 of the programme, based on the government and future NPP owner/operator capabilities and credit worthiness.

---

[3] In this publication, the term 'funding' refers to items that are the responsibility of a government in implementing an NPP (e.g. ensuring the necessary resources for regulation) and the term 'financing' refers to items that are the responsibility of the NPP owner/operator (whether this is the government or a private utility).

The main task of the owner/operator during the FS report preparation is the development of a financial plan for the NPP, a risk analysis of the financing, as well as an economic analysis of the benefits of the project from the viewpoint of the broader society. These aspects will be covered in more detail in the following sections.

The financial plan defines the sources of financing (government budget allocation, local loans, external loans, etc.) and the levels of equity and loans. This plan will also evaluate the cost of financing (IDC, other specific fees, etc.) that will be considered in the economic analyses of the project to determine the optimum solution for the financing.

A successful financial plan also considers the project's susceptibility to potential risks and adverse factors, to acquire the capability to minimize or mitigate their impact, should they materialize. Issues of interest and importance from the viewpoint of financial institutions include the political and economic stability of the nation, the degree of sociopolitical involvement, the prospects of continued economic development, the protection of foreign investment, the promulgation of legislation favourable to nuclear power, the existence of a competent regulatory body and the capability to manage large capital construction projects. A sound financial plan is also a prerequisite to attract NPP vendor interest and allow vendors to bid on the project.

The risk management plan included in the FS report identifies all the key financial risks, their sources, their probability and consequences, and how they are being controlled and mitigated, including the nature of any risk insurance and guarantees. These plans cover the impact of any significant event such as delays in NPP construction, prolonged NPP shutdowns, public liabilities, regulatory delays, political interference and public intervention.

The economic analysis performed within the FS report reassesses the overall costs and benefits of the project to confirm the assumptions made in the nuclear programme initiation within Phase 1. The scope of this analysis may therefore be much broader and account for many elements that contribute to the wellbeing of a society but are not necessarily included in the financial analysis.

3.13.1. Financial analysis

The financial analysis is a detailed analysis of the proposed project from a financial point of view, including a broad range of estimates of the project's costs, a review of the funding requirements and an analysis of the project's financial profitability. Financial analysis ultimately shows whether, with the financing in place, a project is viable and whether the established programmes can bring the project to fruition on time and within budget.

Financial analysis is included in an FS strictly from the point of view of the owner's interests and to facilitate decisions. It therefore begins with analysis of the financial resources of the owner/operator. It would then consider all external financing sources, the type of financing available, or a combination of different sources, such as owner's equity, supplier's credit, buyer's credit, bonds, leasing, joint ventures and institutional investment vehicles such as those provided by international development banks. Any financial analysis would also have to consider the lender's conditions and their requirements and procedures at both the preparatory and implementation phases. The output of this study is the formulation of a number of credible scenarios and a proposal for the financing of the project with due consideration to risks and preconditions.

Inputs to a financial analysis are assumptions on technical, financial, fiscal and economic factors that could impact the financial outcomes of a project. A non-exhaustive list of possible inputs commonly used in evaluating a nuclear power project is provided below:

— Technical assumptions (e.g. plant net capacity, construction period, operational lifetime, start of construction, capacity factor, overnight cost, capital and operational expenses, fuel cost, cost for spent fuel management and decommissioning, radioactive waste management);
— Economic and fiscal assumptions (e.g. inflation, cost escalation factors, exchange rates, taxes, depreciation);

— Financial assumptions (e.g. cost of debt and equity, debt to equity ratio, weighted average cost of capital, upfront cost for debt (bank fee), tenor of debt);
— Contractual arrangements (e.g. EPC turnkey, split contract, fixed price contract or fixed price with escalation, cost based pricing, schedule and conditions of payments);
— Assumptions on electricity markets (e.g. liberalized or regulated market, presence of guaranteed contracts such as power purchase agreements, contracts for difference, feed-in tariffs or other tariffs).

The outputs of the financial analysis are key metrics relevant for the user of the model. These key metrics include net present value (NPV), internal rate of return (IRR), benefit–cost ratio (BCR), levelized cost of electricity (LCOE) or levelized unit cost of energy (LUEC) and different ratios, such as solvency, liquidity, coverage and profitability. More details on these metrics are provided in Section 3.13.3.

3.13.2. Economic analysis

During Phase 1, as part of the energy planning, governments considering the development of NPP projects usually perform an economic analysis to ensure that the overall benefits of the project to a society or an economy exceed its costs, including external and non-monetary costs. The objective of the economic analysis is to help select the project(s) that maximize the benefits for society as a whole and to ensure the optimal allocation of limited financial resources. The main difference between the economic analysis and the financial analysis discussed above is that the economic analysis appraises the project's contribution to the economy as a whole, while financial analysis considers the viewpoint of the owner or investor in the project.

The economic analysis provides indications, solutions and recommendations regarding issues such as:

— The outcome of the project and alternatives;
— The optimal solution to meet the specified objectives;
— The project's financial sustainability;
— The project's environmental impact;
— The impact of the project on the fiscal situation in the country;
— Who will benefit and who will bear the costs of the project;
— The project risks.

During Phase 2, the owner/operator, in coordination with the NEPIO, typically performs an update of the economic analysis as part of the project specific FS with the objective of reconfirming the conclusions of the previous economic analysis or updating them based on more realistic and current inputs. Based on the revised and updated economic analysis, the owner/operator may initiate a review of the policy level project goals and objectives.

The methodology of comparing costs and benefits is in principle similar for an economic or a financial analysis, but the definitions of costs and benefits and the main hypotheses are distinctly different. Overall, the tools of economic analysis can help answer various questions about the project's impact on the entity undertaking the project, on society and on various stakeholders, and about the project's risks and sustainability. Costs and benefits are accounted for as fully as possible, allowing estimation of the net economic benefit associated with the nuclear plant relative to the alternative case where no new investment is undertaken, or investment is made in other technologies.

The cost–benefit analysis is a microeconomic evaluation approach to assessing the impact of a project on society, considering not only the direct economic aspects of the project, but also social and environmental aspects, such as security of supply, emissions of greenhouse gases and other pollutants, and ecological footprint. For these reasons, this analytical tool is commonly used in the public sector to assess an investment decision comprehensively and compare with other alternatives.

3.13.3. Economic and project performance indicators

The most used indicators in financial and economic analysis for project evaluation, as well as the standard metrics for calculating the cost and value of electricity generation, and to capture its value for the system, are described below. Additional financial and economic metrics are described in Ref. [75].

— The NPV of a project is the difference between the present value of all the cash inflows ('positive cashflows') and the present value of all the cash outflows ('negative cashflows') over the lifetime of a project. A positive net present value indicates that the projected earnings generated by a project or investment (in present dollars) exceed the anticipated costs (also in present dollars). Generally, an investment with a positive NPV will be profitable, while one with a negative NPV will result in a net loss. It is also useful for ranking mutually exclusive projects, since the discounted values of alternative cash flows will immediately show which is the more favourable.
— The IRR is the discount rate that makes the net present value of all (i.e. positive and negative) cashflows from a particular investment or project equal to zero. The IRR of a project is typically used to evaluate the attractiveness of a project: if the IRR exceeds the investor's required rate of return, the project is desirable; if it falls below the required rate of return, the project would be rejected.
— The BCR is the ratio of the present value of the lifetime benefit stream to that of the lifetime cost stream. The BCR quantifies the amount of value created by a unit of investment and is therefore useful to rank projects. In general, a project with a BCR lower than 1 is rejected, and the project with the highest BCR is selected. Neither the BCR nor the IRR provide an indication of the actual cashflows.
— The LCOE (sometimes referred to as LUEC) is a standard metric used to compare the electricity generation costs of different power plants. Different power generation technologies have specific characteristics that may vary considerably. The levelized cost methodology allows comparisons between different technologies.

The LCOE represents the average lifetime cost of producing 1 MW·h of electricity, obtained by summing all the various expenses (i.e. investment, fuel, O&M, dismantling and, when appropriate, carbon emissions) over the lifetime of the power plant and dividing them by the electricity generated, after appropriate discounting. These costs are discounted to the commercial operation of an electricity generator.

The LCOE can be viewed as the average price of electricity that equalizes the discounted revenues and expenditures of the plant, making the NPV of the project equal to zero. The lower the LCOE value is, the greater the competitiveness of a plant is.

However, the LCOE lacks representation of the value provided by each plant to the system and hence does not provide guidance on whether a power plant could be competitive in a given electricity system. A better assessment of the economic competitiveness of a power generation technology in a given system can be achieved through joint consideration of the LCOE and other metrics for the power plant value to the grid and its potential revenues for the plant owner/operator. Examples of such metrics are the levelized avoided cost of energy (LACE), developed by the United States Energy Information Administration, or the value adjusted levelized cost of electricity (VALCOE), developed by the IEA and presented in Ref. [75]. The suitability of said metrics for the national electricity market is analysed, taking into account the commercial aspects and technical parameters of the NPP.

3.13.4. Risk assessment

Risk assessment in the context of economic and financial analysis aims at identifying, understanding and examining the project related factors and external events that could impact the forecasted cash flows and revenues of the project. Risk assessment is a fundamental step for all potential providers of financing for any project, including NPP projects. A proper risk assessment is required by investors before deciding to commit their capital to the project and to establish the terms, conditions and price for their investments.

Financial and economic risk assessment [75] involves three broad steps, like the risk management process presented in Section 3.10:

— Risk identification, that is, listing the adverse events to which the project is exposed, identifying the possible causes of the occurrence and the link with the variables and inputs;
— Risk analysis and measuring, that is, determining how assumed changes in the variables determining costs and benefits impact the financial and economic indices calculated (i.e. NPV, IRR or LCOE);
— Development of a risk response plan, risk management and monitoring.

The first step consists of identifying and defining the various risks associated with the project. These risks are of different natures (technical, contractual, legal, financial, reputational, political, market related, etc.), could materialize at different stages of the project (i.e. pre-construction, construction, operation, dismantling and decommissioning) and are controllable or non-controllable by the project developer. A comprehensive list of the project risks (risk register) is created at this stage and will constitute the basis for performing the successive steps of the risk assessment.

In the second step, each of the previously identified risks is analysed individually by assessing the probability of its occurrence and the financial impacts associated with it. A risk matrix, which grades the different risks and their relative impacts, is developed at this stage. The risk matrix allows the ranking and identification of the most critical risks, for which particular monitoring and mitigation efforts are needed.

In the last step, the project developer defines the party (or parties) best suited to managing each project risk during the economic life of the project, identifies the tools or contractual options to mitigate that risk and plans for an efficient risk allocation of the project between the various stakeholders.

The methodology for risk analysis may also include the following tools.

3.13.4.1. Sensitivity analysis

The sensitivity analysis isolates the effect of individual input variables on the overall project performance (usually a selected key performance indicator). Sensitivity analysis is performed to identify what the key input factors influencing an output are and quantify their impact. In practice, a sensitivity analysis is performed by changing, one by one, the assumptions entered into a cost–benefit analysis and seeing the overall impact on the metrics of interest.

Some of the key parameters used for sensitivity analysis in an NPP project are:

— Overnight cost;
— Construction time (delay in the beginning of construction and lead time overrun);
— Electricity market price, selling price or tariff (lower demand or lower market price than expected);
— Discount rate and market interest rates (investment costs higher than expected);
— Exchange rate fluctuation;
— Load factor and plant output (lower than expected);
— Operational costs (fuel costs, O&M costs and labour productivity, maintenance costs);
— Inflation and escalation costs (higher than expected);
— Valuation of externalities (i.e. emissions of greenhouse gases and of other pollutants).

3.13.4.2. Switching values analysis

Similar to sensitivity analysis, switching values analysis determines the percentage by which a variable has to depart from its posted value in order for the net benefits of the project to disappear. Both sensitivity and switching values analysis have two major limitations: they take neither probabilities nor correlations into account. The usual technique of varying one variable at a time is justified only if the variable is uncorrelated with all the other project variables, which most often is not the case.

3.13.4.3. Probabilistic risk analysis

Risk analysis evaluates the probability that a project will (or will not) achieve a satisfactory performance (in terms of NPV, IRR or LUEC), as well as the variance with respect to the best estimate previously made. In contrast to sensitivity analysis, probabilistic risk analysis allows uncertainty in a number of project parameters to be simultaneously accounted for and the impacts on the decision criteria to be quantified. The procedure and specifics of probabilistic risk analysis are presented in Ref. [75].

3.13.4.4. Scenario analysis

A scenario analysis can be undertaken when a meaningful risk analysis cannot be undertaken because of a lack of information on underlying probability distributions or for some other reason. In contrast to sensitivity analysis, which allows analysts to change a single parameter at a time, scenario analysis allows analysts to calculate all of the impacts on the specified indicator of a set of variables at once. 'Optimistic' and 'pessimistic' values of a group of critical variables could be selected to demonstrate different scenarios, within certain hypotheses. However, it is noted that a scenario analysis is just a procedural shortcut and cannot be considered an adequate substitute for a risk analysis.

3.14. GUIDANCE ON OWNER'S SPECIFICATIONS AND REQUIREMENTS

In line with the FS's purpose of serving as a basis for decision making to proceed further with preparation of the contracting phase, the FS includes specific instructions for the approach to developing the owner's specifications and requirements. Depending on the selected contracting approach (i.e. competitive process or sole supplier negotiation) and execution strategy, the instructions in the FS may include the development of a BIS or a document providing the owner's requirements as a basis for the bid selection and negotiation process. Following the approach in Ref. [1], this publication uses the term 'BIS' to describe what is expected to be the subject of this document.

The FS section dedicated to the selection of the EPC contractor ideally includes elaboration of the type of bidding process and its stages, the negotiation strategy and definitions of the main principles of the bid evaluation methodology so that these become part of the decisions to proceed further following FS report approval.

The following subjects, listed in Ref. [2], are to be settled before the vendor selection process begins, or there will at least be plans for their settlement:

— Size or size range of the NPP;
— Feasible NPP designs;
— General technical requirements;
— Applicable codes and standards;
— Experience of potential contractors;
— Overall project schedule;
— Site characteristics;
— Applicable legal frameworks governing nuclear activities;
— Environmental impact;
— Radiation protection;
— Regulatory requirements and licensing procedures;
— Grid characteristics;
— Contractual approaches and project management, including project implementation approach;
— Level of national participation;
— Nuclear fuel supply options;
— Nuclear waste management and disposal;

— Nuclear safety and nuclear safety culture requirements;
— Security and physical protection;
— Safeguards;
— Economics;
— Financing options;
— Quality assurance and control requirements;
— Owner/operator's scope of supply;
— Owner/operator's organization;
— Emergency planning.

The above elements are expected to be available or to be derived from the various analyses performed in the framework of the FS development. Therefore, it is important to identify which elements are taken from the FS efforts, to ensure that the subsequent vendor selection process will be carried out within the project execution modelling assessed by the FS, and which ones would have to be additionally or subsequently developed as part of the BIS development process.

Other elements that need to be defined by the owner/operator and estimated in this section of the FS are the human resources required to prepare the BIS and subsequently to conduct the bid evaluation, as well as the expected sequencing and period of this process. Specific human resources needs may include:

— Technical expertise to develop the specifications for an NPP and evaluate the bids;
— Project and management system expertise to manage the bidding process and to develop specifications and evaluate the bids;
— Detailed knowledge of the infrastructure in the country and at the site, as well as the regulatory environment;
— Legal and business expertise for the BIS preparation, bid evaluation, contract negotiations and fuel procurement;
— Financing expertise to negotiate with financing organizations and develop financing plans;
— Expertise in stakeholder communication and public information.

More details about the functions and responsibilities of the owner/operator in preparation of the BIS and evaluation of the bids are presented in Ref. [1] and need to be studied during the FS preparation to ensure that sufficient resources will be included in the estimates.

As discussed in Ref. [2], the bidding process can be divided into the following main phases:

— Preparation of the BIS;
— Preparation of bids;
— Evaluation of bids;
— Contract negotiation and signature.

Typically, the bid evaluation process will include the following steps:

— Preliminary bid evaluation;
— Preparation of preliminary bid evaluation report;
— Deciding on the shortlist of bidders, if feasible;
— Detailed bid evaluation;
— Preparation of final evaluation report.

The bidding process typically spans 20–40 months [2]. Therefore, it is important that assumptions are elaborated in this section of the FS with respect to the sequence, procedure and duration of the bidding process to plan for BIS preparation and issue.

4. CONCLUSIONS

An FS is an analysis whose primary purpose is to support owners and other stakeholders to reach the correct business decisions. It is also written in such a way that it will be useful as an authoritative record of the reasoning and methodologies employed to support the decision to build an NPP, including risk analysis and legal implications, as well as a reference document that can be used as a basis for further studies, for the preparation of the BIS and for the information package requested by prospective financial institutions, export banks and private investor groups.

In summary, key risk elements need to be addressed in an FS for an NPP as follows:

(a) Various contractual approaches are examined impartially, together with the possible procurement models and the project management styles before recommendations are made.

(b) Various partnership models with a variety of capital injections are examined. In addition, various contractual approaches are examined, together with the procurement models and the project management styles, and recommendations are made.

(c) It is also important to describe the accident preparedness and site emergency planning, the infrastructure required, if any, and its cost. Various options are compared before a solution is recommended.

(d) Before a final recommendation for unit capacity is made, the risks that may be undertaken when selecting larger unit sizes are well assessed, together with the risks associated with grid size, while all other aspects are also carefully evaluated.

(e) Recommendations on the reactor type and size are preceded by an accurate technology market survey, and the assessment process is in strict adherence to the programme policy objectives, the NPP project goal, the waste management objectives, the fuel cycle evaluation, a comprehensive safety assessment based on international experience and the recommendations from the EIA.

(f) If the regulatory framework and licensing requirements have not been completed or enforced, it is expected that this section of the FS report will be prepared based solely on the reference information received from the potential NPP vendor countries, including any input from the regulatory bodies of these countries.

(g) A detailed national industry detailed survey, the establishment of a strategy for national participation, a localization plan and a technology transfer plan (if required) are conducted to a degree that allows for further, more specific, studies and for the BIS.

(h) In the cost estimate for the project, in addition to financial costs, any social costs or benefits other than compensation flowing from the project to third parties are accounted for, quantified by assigning realistic monetary values.

A project risk matrix and its implications, followed by a risk management plan, are developed and included in the FS.

Given the complexity and multidisciplinary characteristics of an FS for an NPP project, it is recommended that a consulting firm or TSO be used with solid experience in creating successful FSs, with qualified key personnel with proven knowledge of the nuclear industry and of the variety of topics making up the study, and a good knowledge and understanding of the cultural environment and of the owner/operator organization and its interfaces.

REFERENCES

[1] INTERNATIONAL ATOMIC ENERGY AGENCY, Milestones in the Development of a National Infrastructure for Nuclear Power, IAEA Nuclear Energy Series No. NG-G-3.1 (Rev. 2), IAEA, Vienna (2024), https://doi.org/10.61092/iaea.zjau-e8cs

[2] INTERNATIONAL ATOMIC ENERGY AGENCY, Invitation and Evaluation of Bids for Nuclear Power Plants, IAEA Nuclear Energy Series No. NG-T-3.9, IAEA, Vienna (2011).

[3] INTERNATIONAL ATOMIC ENERGY AGENCY, Initiating Nuclear Power Programmes: Responsibilities and Capabilities of Owners and Operators, IAEA Nuclear Energy Series No. NG-T-3.1 (Rev. 1), IAEA, Vienna (2020).

[4] INTERNATIONAL ATOMIC ENERGY AGENCY, Responsibilities and Functions of a Nuclear Energy Programme Implementing Organization, IAEA Nuclear Energy Series No. NG-T-3.6 (Rev. 1), IAEA, Vienna (2019).

[5] INTERNATIONAL ATOMIC ENERGY AGENCY, Technical Support to Nuclear Power Plants and Programmes, IAEA Nuclear Energy Series No. NP-T-3.28, IAEA, Vienna (2018).

[6] INTERNATIONAL ATOMIC ENERGY AGENCY, Stakeholder Engagement in Nuclear Programmes, IAEA Nuclear Energy Series No. NG-G-5.1, IAEA, Vienna (2021).

[7] INTERNATIONAL ATOMIC ENERGY AGENCY, Governmental, Legal and Regulatory Framework for Safety, IAEA Safety Standards Series No. GSR Part 1 (Rev. 1), IAEA, Vienna (2016).

[8] INTERNATIONAL ATOMIC ENERGY AGENCY, Licensing Process for Nuclear Installations, IAEA Safety Standards Series No. SSG-12, IAEA, Vienna (2010).

[9] INTERNATIONAL ATOMIC ENERGY AGENCY, Model for Analysis of Energy Demand (MAED-2), Computer Manual Series No. 18, IAEA, Vienna (2006).

[10] INTERNATIONAL ATOMIC ENERGY AGENCY, Electric Grid Reliability and Interface with Nuclear Power Plants, IAEA Nuclear Energy Series No. NG-T-3.8, IAEA, Vienna (2012).

[11] INTERNATIONAL ATOMIC ENERGY AGENCY, Nuclear Reactor Technology Assessment for Near Term Deployment, IAEA Nuclear Energy Series No. NP-T-1.10, IAEA, Vienna (2013).

[12] INTERNATIONAL ATOMIC ENERGY AGENCY, Evaluation of the Status of National Nuclear Infrastructure Development, IAEA Nuclear Energy Series No. NG-T-3.2 (Rev. 2), IAEA, Vienna (2022).

[13] INTERNATIONAL ATOMIC ENERGY AGENCY, Managing Siting Activities for Nuclear Power Plants, IAEA Nuclear Energy Series No. NG-T-3.7 (Rev. 1), IAEA, Vienna (2022).

[14] INTERNATIONAL ATOMIC ENERGY AGENCY, Hazards Associated with Human Induced External Events in Site Evaluation for Nuclear Installations, IAEA Safety Standards Series No. SSG-79, IAEA, Vienna (2023).

[15] INTERNATIONAL ATOMIC ENERGY AGENCY, Geotechnical Aspects of Site Evaluation and Foundations for Nuclear Power Plants, IAEA Safety Standards Series No. NS-G-3.6, IAEA, Vienna (2005).

[16] INTERNATIONAL ATOMIC ENERGY AGENCY, Volcanic Hazards in Site Evaluation for Nuclear Installations, IAEA Safety Standards Series No. SSG-21, IAEA, Vienna (2012).

[17] INTERNATIONAL ATOMIC ENERGY AGENCY, Seismic Hazards in Site Evaluation for Nuclear Installations, IAEA Safety Standards Series No. SSG-9 (Rev. 1), IAEA, Vienna (2022).

[18] INTERNATIONAL ATOMIC ENERGY AGENCY, Advanced Nuclear Power Plant Design Options to Cope with External Events, IAEA-TECDOC-1487, IAEA, Vienna (2006).

[19] INTERNATIONAL ATOMIC ENERGY AGENCY, Site Survey and Site Selection for Nuclear Installations, IAEA Safety Standards Series No. SSG-35, IAEA, Vienna (2015).

[20] INTERNATIONAL ATOMIC ENERGY AGENCY, Dispersion of Radioactive Material in Air and Water and Consideration of Population Distribution in Site Evaluation for Nuclear Power Plants, IAEA Safety Standards Series No. NS-G-3.2, IAEA, Vienna (2002).

[21] INTERNATIONAL ATOMIC ENERGY AGENCY, Meteorological and Hydrological Hazards in Site Evaluation for Nuclear Installations, IAEA Safety Standards Series No. SSG-18, Vienna (2011).

[22] INTERNATIONAL ATOMIC ENERGY AGENCY, Disposal of Radioactive Waste, IAEA Safety Standards Series No. SSR-5, IAEA, Vienna (2011).

[23] INTERNATIONAL ATOMIC ENERGY AGENCY, Considerations in the Development of Near Surface Repositories for Radioactive Waste, Technical Reports Series No. 417, IAEA, Vienna (2003).

[24] INTERNATIONAL ATOMIC ENERGY AGENCY, Hydrogeological Investigation of Sites for the Geological Disposal of Radioactive Waste, Technical Reports Series No. 391, IAEA, Vienna (1999).

[25] INTERNATIONAL ATOMIC ENERGY AGENCY, Safety Guide on Siting of Near Surface Disposal Facilities, IAEA Safety Standards Series No. 111-G-3.1, IAEA, Vienna (1994).

[26] INTERNATIONAL ATOMIC ENERGY AGENCY, Geological Disposal Facilities for Radioactive Waste, IAEA Safety Standards Series No. SSG-14, IAEA, Vienna (2011).

[27] INTERNATIONAL ATOMIC ENERGY AGENCY, Site Evaluation for Nuclear Installations, IAEA Safety Standards Series No. SSR-1, IAEA, Vienna (2019).

[28] INTERNATIONAL ATOMIC ENERGY AGENCY, Nuclear Energy Basic Principles, IAEA Nuclear Energy Series No. NE-BP, IAEA, Vienna (2008).

[29] INTERNATIONAL ATOMIC ENERGY AGENCY, Strategic Environmental Assessment for Nuclear Power Programmes: Guidelines, IAEA Nuclear Energy Series No. NG-T-3.17, IAEA, Vienna (2018).

[30] INTERNATIONAL ATOMIC ENERGY AGENCY, Environmental Protection in New Nuclear Power Programmes, IAEA Nuclear Energy Series No. NG-T-3.11 (Rev. 1), IAEA, Vienna (2024).

[31] INTERNATIONAL ATOMIC ENERGY AGENCY, Advanced Reactor Information System (ARIS), http://aris.iaea.org

[32] INTERNATIONAL ATOMIC ENERGY AGENCY, Survey of Wet and Dry Spent Fuel Storage, IAEA-TECDOC-1100, IAEA, Vienna (1999).

[33] INTERNATIONAL ATOMIC ENERGY AGENCY, Behaviour of Spent Power Reactor Fuel During Storage, IAEA-TECDOC-1862, IAEA, Vienna (2019).

[34] INTERNATIONAL ATOMIC ENERGY AGENCY, Storing Spent Fuel until Transport to Reprocessing or Disposal, IAEA Nuclear Energy Series No. NF-T-3.3, IAEA, Vienna (2019).

[35] INTERNATIONAL ATOMIC ENERGY AGENCY, Options for Management of Spent Nuclear Fuel and Radioactive Waste for Countries Developing New Nuclear Power Programmes, IAEA Nuclear Energy Series No. NW-T-1.24 (Rev. 1), IAEA, Vienna (2018).

[36] INTERNATIONAL ATOMIC ENERGY AGENCY, Costing Methods and Funding Schemes for Radioactive Waste Disposal Programmes, IAEA Nuclear Energy Series No. NW-T-1.25, IAEA, Vienna (2020).

[37] INTERNATIONAL ATOMIC ENERGY AGENCY, Management of Nuclear Power Plant Projects, IAEA Nuclear Energy Series No. NG-T-1.6, IAEA, Vienna (2020).

[38] INTERNATIONAL ENERGY AGENCY, OECD NUCLEAR ENERGY AGENCY, Projected Costs of Electricity Generation, 2020 Edition, OECD, Paris (2020).

[39] INTERNATIONAL ATOMIC ENERGY AGENCY, Economic Evaluation of Bids for Nuclear Power Plants, Technical Reports Series No. 396, IAEA, Vienna (2000).

[40] D'HAESELEER, W.D., Synthesis of the Economics of Nuclear Energy, Study for the European Commission, EC, Leuven (2013).

[41] OECD NUCLEAR ENERGY AGENCY, The Economics of the Back End of the Nuclear Fuel Cycle, OECD, Paris (2013).

[42] INTERNATIONAL ENERGY AGENCY, OECD NUCLEAR ENERGY AGENCY, Projected Costs of Electricity Generation, 2010 Edition, OECD, Paris (2010).

[43] INTERNATIONAL ENERGY AGENCY, OECD NUCLEAR ENERGY AGENCY, Projected Costs of Electricity Generation, 2015 Edition, OECD, Paris (2015).

[44] INTERNATIONAL ATOMIC ENERGY AGENCY, Project Management in Nuclear Power Plant Construction: Guidelines and Experience, IAEA Nuclear Energy Series No. NP-T-2.7, IAEA, Vienna (2012).

[45] INTERNATIONAL ATOMIC ENERGY AGENCY, Managing the Financial Risk Associated with the Financing of New Nuclear Power Plant Projects, IAEA Nuclear Energy Series No. NG-T-4.6, IAEA, Vienna (2017).

[46] INTERNATIONAL ATOMIC ENERGY AGENCY, Contracting and Ownership Approaches for New Nuclear Power Plants, IAEA-TECDOC-1750/Rev. 1, IAEA, Vienna (2024).

[47] INTERNATIONAL ATOMIC ENERGY AGENCY, The Nuclear Contracting Toolkit, https://nucleus.iaea.org/sites/connect/MSNpublic/Pages/nct/site/index.html

[48] INTERNATIONAL ATOMIC ENERGY AGENCY, Fundamental Safety Principles, IAEA Safety Standards Series No. SF-1, IAEA, Vienna (2006), https://doi.org/10.61092/iaea.hmxn-vw0a

[49] INTERNATIONAL ATOMIC ENERGY AGENCY, Safety of Nuclear Power Plants: Design, IAEA Safety Standards Series No. SSR-2/1 (Rev. 1), IAEA, Vienna (2016).

[50] INTERNATIONAL ATOMIC ENERGY AGENCY, Design of Fuel Handling and Storage Systems for Nuclear Power Plants, IAEA Safety Standards Series No. SSG-63, IAEA, Vienna (2020).

[51] INTERNATIONAL ATOMIC ENERGY AGENCY, Design of Nuclear Installations Against External Events Excluding Earthquakes, IAEA Safety Standards Series No. SSG-68, IAEA, Vienna (2021).

[52] INTERNATIONAL ATOMIC ENERGY AGENCY, Seismic Design for Nuclear Installations, IAEA Safety Standards Series No. SSG-67, IAEA, Vienna (2021).

[53] INTERNATIONAL ATOMIC ENERGY AGENCY, Protection Against Internal Hazards in the Design of Nuclear Power Plants, IAEA Safety Standards Series No. SSG-64, IAEA, Vienna (2021).

[54] INTERNATIONAL ATOMIC ENERGY AGENCY, Design of the Reactor Coolant System and Associated Systems for Nuclear Power Plants, IAEA Safety Standards Series No. SSG-56, IAEA, Vienna (2020).

[55] INTERNATIONAL ATOMIC ENERGY AGENCY, Protection Against Internal Hazards in the Design of Nuclear Power Plants, IAEA Safety Standards Series No. SSG-64, IAEA, Vienna (2021).

[56] INTERNATIONAL ATOMIC ENERGY AGENCY, Radiation Protection Aspects of Design for Nuclear Power Plants, IAEA Safety Standards Series No. NS-G-1.13, IAEA, Vienna (2005).

[57] INTERNATIONAL ATOMIC ENERGY AGENCY, Safety Classification of Structures, Systems and Components in Nuclear Power Plants, IAEA Safety Standards Series No. SSG-30, IAEA, Vienna (2014).

[58] INTERNATIONAL ATOMIC ENERGY AGENCY, Design of Electrical Power Systems for Nuclear Power Plants, IAEA Safety Standards Series No. SSG-34, IAEA, Vienna (2016).

[59] INTERNATIONAL ATOMIC ENERGY AGENCY, Design of Instrumentation and Control Systems for Nuclear Power Plants, IAEA Safety Standards Series No. SSG-39, IAEA, Vienna (2016).

[60] INTERNATIONAL ATOMIC ENERGY AGENCY, Design of the Reactor Core for Nuclear Power Plants, IAEA Safety Standards Series No. SSG-52, IAEA, Vienna (2019).

[61] INTERNATIONAL ATOMIC ENERGY AGENCY, Design of the Reactor Containment and Associated Systems for Nuclear Power Plants, IAEA Safety Standards Series No. SSG-53, IAEA, Vienna (2019).

[62] INTERNATIONAL NUCLEAR SAFETY ADVISORY GROUP, Maintaining the Design Integrity of Nuclear Installations Throughout Their Operating Life, INSAG Series No. 19, IAEA, Vienna (2003).

[63] INTERNATIONAL ATOMIC ENERGY AGENCY, Construction Technologies for Nuclear Power Plants, IAEA Nuclear Energy Series No. NP-T-2.5, IAEA, Vienna (2011).

[64] UNITED STATES DEPARTMENT OF ENERGY, Application of Advanced Construction Technologies to New Nuclear Power Plants, MPR 2610 Revision 2, USDOE, Washington, DC (2004).

[65] INTERNATIONAL ATOMIC ENERGY AGENCY, Construction for Nuclear Installations, IAEA Safety Standards Series No. SSG-38, IAEA, Vienna (2015).

[66] INTERNATIONAL ATOMIC ENERGY AGENCY, Commissioning for Nuclear Power Plants, IAEA Safety Standards Series No. SSG-28, IAEA, Vienna (2014).

[67] INTERNATIONAL ATOMIC ENERGY AGENCY, Commissioning of Nuclear Power Plants: Training and Human Resource Considerations, IAEA Nuclear Energy Series No. NG-T-2.2, IAEA, Vienna (2008).

[68] INTERNATIONAL ATOMIC ENERGY AGENCY, Commissioning Guidelines for Nuclear Power Plants, IAEA Nuclear Energy Series No. NP-T-2.10, IAEA, Vienna (2018).

[69] INTERNATIONAL ATOMIC ENERGY AGENCY, Quality Assurance and Quality Control in Nuclear Facilities and Activities, IAEA-TECDOC-1910, IAEA, Vienna (2020).

[70] INTERNATIONAL ATOMIC ENERGY AGENCY, Integrated Life Cycle Risk Management for New Nuclear Power Plants, IAEA Nuclear Energy Series No. NR-T-2.15, IAEA, Vienna (2023).

[71] INTERNATIONAL ATOMIC ENERGY AGENCY, Risk Management: A Tool for Improving Nuclear Power Plant Performance, IAEA-TECDOC-1209, IAEA, Vienna (2001).

[72] INTERNATIONAL ATOMIC ENERGY AGENCY, Human Resource Management for New Nuclear Power Programmes, IAEA Nuclear Energy Series No. NG-T-3.10 (Rev. 1), IAEA, Vienna (2022).

[73] INTERNATIONAL ATOMIC ENERGY AGENCY, Staffing Requirements for Future Small and Medium Reactors (SMRs) Based on Operating Experience and Projections, IAEA-TECDOC-1193, IAEA, Vienna (2001).

[74] INTERNATIONAL ATOMIC ENERGY AGENCY, Industrial Involvement to Support a Programme of Nuclear Power, IAEA Nuclear Energy Series No. NG-T-3.4, IAEA, Vienna (2016).

[75] INTERNATIONAL ATOMIC ENERGY AGENCY, Financing Nuclear Power Plants, IAEA-TECDOC-1964, IAEA, Vienna (2021).

ABBREVIATIONS

ARIS	Advanced Reactor Information System
BCR	benefit–cost ratio
BIS	bid invitation specification
EIA	environmental impact assessment
EMP	environmental management plan
EPC	engineering, procurement and construction
ESR	environmental scoping report
FS	feasibility study
IDC	interest during construction
IRR	internal rate of return
LCOE	levelized cost of electricity
LUEC	levelized unit cost of energy
MWI	megawatt installed
NEPIO	nuclear energy programme implementing organization
NPP	nuclear power plant
NPV	net present value
O&M	operation and maintenance
pre-FS	pre-feasibility study
RFI	request for information
RTA	reactor technology assessment
TCIC	total capital investment cost
TSO	technical support organization

CONTRIBUTORS TO DRAFTING AND REVIEW

Structure of the IAEA Nuclear Energy Series*

Nuclear Energy Basic Principles
NE-BP

Nuclear Energy General Objectives
NG-O

1. Management Systems
NG-G-1.#
NG-T-1.#

2. Human Resources
NG-G-2.#
NG-T-2.#

3. Nuclear Infrastructure and Planning
NG-G-3.#
NG-T-3.#

4. Economics and Energy System Analysis
NG-G-4.#
NG-T-4.#

5. Stakeholder Involvement
NG-G-5.#
NG-T-5.#

6. Knowledge Management
NG-G-6.#
NG-T-6.#

Nuclear Reactor Objectives**
NR-O

1. Technology Development
NR-G-1.#
NR-T-1.#

2. Design, Construction and Commissioning of Nuclear Power Plants
NR-G-2.#
NR-T-2.#

3. Operation of Nuclear Power Plants
NR-G-3.#
NR-T-3.#

4. Non Electrical Applications
NR-G-4.#
NR-T-4.#

5. Research Reactors
NR-G-5.#
NR-T-5.#

Nuclear Fuel Cycle Objectives
NF-O

1. Exploration and Production of Raw Materials for Nuclear Energy
NF-G-1.#
NF-T-1.#

2. Fuel Engineering and Performance
NF-G-2.#
NF-T-2.#

3. Spent Fuel Management
NF-G-3.#
NF-T-3.#

4. Fuel Cycle Options
NF-G-4.#
NF-T-4.#

5. Nuclear Fuel Cycle Facilities
NF-G-5.#
NF-T-5.#

Radioactive Waste Management and Decommissioning Objectives
NW-O

1. Radioactive Waste Management
NW-G-1.#
NW-T-1.#

2. Decommissioning of Nuclear Facilities
NW-G-2.#
NW-T-2.#

3. Environmental Remediation
NW-G-3.#
NW-T-3.#

(*) as of 1 January 2020
(**) Formerly 'Nuclear Power' (NP)

Key
BP: Basic Principles
O: Objectives
G: Guides and Methodologies
T: Technical Reports
Nos 1–6: Topic designations
#: Guide or Report number

Examples
NG-G-3.1: Nuclear Energy General (**NG**), Guides and Methodologies (**G**), Nuclear Infrastructure and Planning (topic **3**), **#1**
NR-T-5.4: Nuclear Reactors (**NR**), Technical Report (**T**), Research Reactors (topic **5**), **#4**
NF-T-3.6: Nuclear Fuel (**NF**), Technical Report (**T**), Spent Fuel Management (topic **3**), **#6**
NW-G-1.1: Radioactive Waste Management and Decommissioning (**NW**), Guides and Methodologies (**G**), Radioactive Waste Management (topic **1**) **#1**

CONTACT IAEA PUBLISHING

Feedback on IAEA publications may be given via the on-line form available at:
www.iaea.org/publications/feedback

This form may also be used to report safety issues or environmental queries concerning IAEA publications.

Alternatively, contact IAEA Publishing:

Publishing Section
International Atomic Energy Agency
Vienna International Centre, PO Box 100, 1400 Vienna, Austria
Telephone: +43 1 2600 22529 or 22530
Email: sales.publications@iaea.org
www.iaea.org/publications

Priced and unpriced IAEA publications may be ordered directly from the IAEA.

ORDERING LOCALLY

Priced IAEA publications may be purchased from regional distributors and from major local booksellers.

Printed and bound by CPI Group (UK) Ltd, Croydon, CR0 4YY

06/07/2026

02160600-0003